Food Carbohydrate Chemistry

Press

The *IFT Press* series reflects the mission of the Institute of Food Technologists—to advance the science of food contributing to healthier people everywhere. Developed in partnership with Wiley-Blackwell, *IFT Press* books serve as leading-edge handbooks for industrial application and reference and as essential texts for academic programs. Crafted through rigorous peer review and meticulous research, *IFT Press* publications represent the latest, most significant resources available to food scientists and related agriculture professionals worldwide. Founded in 1939, the Institute of Food Technologists is a nonprofit scientific society with 22,000 individual members working in food science, food technology, and related professions in industry, academia, and government. IFT serves as a conduit for multidisciplinary science thought leadership, championing the use of sound science across the food value chain through knowledge sharing, education, and advocacy.

A John Wiley & Sons, Ltd., Publication

Food Carbohydrate Chemistry

Ronald E. Wrolstad
Distinguished Professor of Food Science Emeritus
Oregon State University

A John Wiley & Sons, Ltd., Publication

This edition first published 2012 © 2012 by John Wiley & Sons, Inc. and Institute of Food Technologists

Wiley-Blackwell is an imprint of John Wiley & Sons, formed by the merger of Wiley's global Scientific, Technical and Medical business with Blackwell Publishing.

Registered office: John Wiley & Sons Ltd, The Atrium, Southern Gate, Chichester, West Sussex, PO19 8SQ, UK

Editorial offices: 2121 State Avenue, Ames, Iowa 50014-8300, USA
The Atrium, Southern Gate, Chichester, West Sussex, PO19 8SQ, UK
9600 Garsington Road, Oxford, OX4 2DQ, UK

For details of our global editorial offices, for customer services and for information about how to apply for permission to reuse the copyright material in this book please see our website at www.wiley.com/wiley-blackwell.

Library of Congress Cataloging-in-Publication Data

Wrolstad, Ronald E., 1939–
  Food carbohydrate chemistry / Ronald E. Wrolstad. – 1st ed.
    p. cm. – (Institute of food technologists series ; 48)
  Includes bibliographical references and index.
  ISBN 978-0-8138-2665-3 (hardback)
1. Carbohydrates.   I. Title.
  QD321.W88 2012
  547'.78–dc23
                                        2011036449

A catalogue record for this book is available from the British Library.

Wiley also publishes its books in a variety of electronic formats. Some content that appears in print may not be available in electronic books.

Set in 9.5/12.5pt Palatino by Aptara® Inc., New Delhi, India
Printed and bound in Malaysia by Vivar Printing Sdn Bhd

1   2012

Titles in the IFT Press series

- *Accelerating New Food Product Design and Development* (Jacqueline H. Beckley, Elizabeth J. Topp, M. Michele Foley, J.C. Huang, and Witoon Prinyawiwatkul)
- *Advances in Dairy Ingredients* (Geoffrey W. Smithers and Mary Ann Augustin)
- *Bioactive Proteins and Peptides as Functional Foods and Nutraceuticals* (Yoshinori Mine, Eunice Li–Chan, and Bo Jiang)
- *Biofilms in the Food Environment* (Hans P. Blaschek, Hua H. Wang, and Meredith E. Agle)
- *Calorimetry in Food Processing: Analysis and Design of Food Systems* (Gönül Kaletunç)
- *Coffee: Emerging Health Effects and Disease Prevention* (YiFang Chu)
- *Food Carbohydrate Chemistry* (Ronald E. Wrolstad)
- *Food Ingredients for the Global Market* (Yao–Wen Huang and Claire L. Kruger)
- *Food Irradiation Research and Technology* (Christopher H. Sommers and Xuetong Fan)
- *Foodborne Pathogens in the Food Processing Environment: Sources, Detection and Control* (Sadhana Ravishankar, Vijay K. Juneja, and Divya Jaroni)
- *High Pressure Processing of Foods* (Christopher J. Doona and Florence E. Feeherry)
- *Hydrocolloids in Food Processing* (Thomas R. Laaman)
- *Improving Import Food Safety* (Wayne C. Ellefson, Lorna Zach, and Darryl Sullivan)
- *Innovative Food Processing Technologies: Advances in Multiphysics Simulation* (Kai Knoerzer, Pablo Juliano, Peter Roupas, and Cornelis Versteeg)
- *Microbial Safety of Fresh Produce* (Xuetong Fan, Brendan A. Niemira, Christopher J. Doona, Florence E. Feeherry, and Robert B. Gravani)
- *Microbiology and Technology of Fermented Foods* (Robert W. Hutkins)
- *Multiphysics Simulation of Emerging Food Processing Technologies* (Kai Knoerzer, Pablo Juliano, Peter Roupas and Cornelis Versteeg)
- *Multivariate and Probabilistic Analyses of Sensory Science Problems* (Jean–François Meullenet, Rui Xiong, and Christopher J. Findlay
- *Nanoscience and Nanotechnology in Food Systems* (Hongda Chen)
- *Natural Food Flavors and Colorants* (Mathew Attokaran)
- *Nondestructive Testing of Food Quality* (Joseph Irudayaraj and Christoph Reh)
- *Nondigestible Carbohydrates and Digestive Health* (Teresa M. Paeschke and William R. Aimutis)
- *Nonthermal Processing Technologies for Food* (Howard Q. Zhang, Gustavo V. Barbosa–Cánovas, V.M. Balasubramaniam, C. Patrick Dunne, Daniel F. Farkas, and James T.C. Yuan)
- *Nutraceuticals, Glycemic Health and Type 2 Diabetes* (Vijai K. Pasupuleti and James W. Anderson)
- *Organic Meat Production and Processing* (Steven C. Ricke, Michael G. Johnson, and Corliss A. O'Bryan)
- *Packaging for Nonthermal Processing of Food* (Jung H. Han)
- *Preharvest and Postharvest Food Safety: Contemporary Issues and Future Directions* (Ross C. Beier, Suresh D. Pillai, and Timothy D. Phillips, Editors; Richard L. Ziprin, Associate Editor)
- *Processing and Nutrition of Fats and Oils* (Ernesto M. Hernandez and Afaf Kamal–Eldin)
- *Processing Organic Foods for the Global Market* (Gwendolyn V. Wyard, Anne Plotto, Jessica Walden, and Kathryn Schuett)
- *Regulation of Functional Foods and Nutraceuticals: A Global Perspective* (Clare M. Hasler)
- *Resistant Starch: Sources, Applications and Health Benefits* (Yong–Cheng Shi and Clodualdo Maningat)

- *Sensory and Consumer Research in Food Product Design and Development* (Howard R. Moskowitz, Jacqueline H. Beckley, and Anna V.A. Resurreccion)
- *Sustainability in the Food Industry* (Cheryl J. Baldwin)
- *Thermal Processing of Foods: Control and Automation* (K.P. Sandeep)
- *Trait–Modified Oils in Foods* (Frank T. Orthoefer and Gary R. List)
- *Water Activity in Foods: Fundamentals and Applications* (Gustavo V. Barbosa–Cánovas, Anthony J. Fontana Jr., Shelly J. Schmidt, and Theodore P. Labuza)
- *Whey Processing, Functionality and Health Benefits* (Charles I. Onwulata and Peter J. Huth)

Dedication

This book is dedicated to two special mentors, one being my Major Professor at the University of California, Davis, Dr. Walter G. Jennings. His concern for students and his enthusiasm for research and teaching continue to inspire. The second is the late Robert S. Shallenberger with whom I was fortunate to work while on sabbatical leave at Cornell University in 1979–1980. His influence on this book should be evident on nearly every page. I would also like to dedicate the book to the many undergraduate and graduate students, who through their suggestions, understanding, and misunderstanding helped me to revise, discard, and improve lecture presentations, homework assignments, demonstrations, and laboratory exercises. All of those items were a platform for this book.

Contents

Contributors

Chapter 7

Andrew S. Ross
Department of Crop and Soil Science/Department of Food Science and Technology
Oregon State University
Corvallis, Oregon

Chapter 8

Bronwen G. Smith and Laurence D. Melton
Food Science Programme
The University of Auckland
Auckland, New Zealand

Acknowledgments

A sincere thanks to Andrew Ross, who authored Chapter 7, and to Laurence Melton and Bronwen Smith for Chapter 8. Thanks also to Dan Smith for his insightful reviewing and to Carole Jubert, who came to the rescue of this novice in ChemDraw^{TM} and prepared the chemical structures and figures.

Introduction

Carbohydrates are major components of foods, accounting for more than 90% of the dry matter of fruits and vegetables and providing for 70–80% of human caloric intake worldwide (BeMiller and Huber 2008). Thus, from a quantitative perspective alone, carbohydrates warrant the attention of food chemists. From the standpoint of food quality, carbohydrates are multifunctional. Sugars are the major source, as well as our reference point, for sweetness. Although carbohydrates are described as being odorless, the volatile reaction products from the Maillard reaction, Strecker degradation, and carmelization reactions can provide desirable, undesirable, or neutral flavor compounds. And, although carbohydrates are colorless, sugars participate in Maillard and carmelization reactions to produce desirable and undesirable brown colors. Cellulose, hemicellulose, pectin, and starch are the structural components of plants that are largely responsible for the textural characteristics of fruits and vegetables. Starch and starch derivatives and various hydrocolloids isolated from plants, seaweed, and microbial sources are used as thickeners, gelling agents, bodying agents, and stabilizers in foods. When it comes to nutrition, a sizable portion of the lay public view carbohydrates in a bad light. Carbohydrates are often blamed for health issues such as obesity, diabetes, and dental caries. It should be realized that carbohydrates are, or should be, the principal source of energy in our diet. After all, we evolved as a species to efficiently use carbohydrates that can be converted to glucose for our body's fuel. Good nutrition is based on the consumption of the appropriate carbohydrates in the right amounts in balance with other nutrients. It is widely accepted that consumption of various forms of complex carbohydrate can reduce the risk of diabetes, coronary heart disease, diverticulitus, and colon cancer. For peak athletic performance, the advice of professional nutritionists will emphasize consumption of the appropriate carbohydrates, in the appropriate amounts, at the appropriate time. Although the percentage of carbohydrates contributing to caloric intake in the United States is highly variable, the average is considerably less than 70%. Dietary recommendations call for increased consumption of fruits and vegetables and a greater proportion of complex carbohydrate (Walker and Reamy 2009; WHO 2010).

The major thrust of this book is to apply basic carbohydrate chemistry to the quality attributes and functional properties of foods. Structure and nomenclature of sugars and sugar derivatives is covered but limited to those compounds that exist naturally in foods or are used as food additives and food ingredients. Review and presentation of fundamental carbohydrate chemistry is minimized, with the assumption that readers have taken general organic chemistry and general biochemistry and have ready access to those books for reference. Chemical reactions focus on those that have an impact on food quality and occur under processing and storage conditions. How chemical and physical properties of sugars and polysaccharides affect the functional properties of foods is emphasized. Taste properties and nonenzymic browning reactions are covered. The nutritional roles of carbohydrates are covered from a food chemist's perspective. One chapter describes selected carbohydrate analytical methods, emphasizing the basic principles of the methods and their advantages and limitations. There is an extensive appendix that includes some laboratory and classroom exercises and lecture demonstrations.

References

BeMiller JM, Huber KC. 2008. Carbohydrates. In: Damodaran S, Parkin KL, Fennema OR, editors. *Fennema's Food Chemistry, 4th ed.* Boca Raton, FL: CRC Press, Taylor & Francis, pp. 83–154.

Walker C, Reamy BV. 2009. Diets for cardiovascular disease prevention: what is the evidence? *Am Fam Physician* 79:571–8. Available from: http://www.ncbi.nlm.nih.gov/pubmed/19378874. Accessed September 2, 2010.

WHO 2010. Global strategy on diet, physical activity and health. Available from: http://www.who.int/dietphysicalactivity/diet/en/index.html. Accessed September 2, 2010.

1 Classifying, Identifying, Naming, and Drawing Sugars and Sugar Derivatives

Food Carbohydrate Chemistry, First Edition. Ronald E. Wrolstad.
© 2012 John Wiley & Sons, Inc. Published 2012 by John Wiley & Sons, Inc.

Structure and Nomenclature of Monosaccharides

Sugars are polyhydroxycarbonyls that occur in single or multiple units as monosaccharides, disaccharides, trisaccharides, tetrasaccharides, or oligosaccharides (typically three to ten sugar units). Monosaccharides (also known as simple sugars) exist as aldoses or ketoses, with glucose and fructose being the most common examples. **Glycose** is a generic term for sugars. Sugars are also classified according to the number of carbon atoms in the molecule (e.g., trioses, tetroses, pentoses, hexoses, heptoses, etc.).

Aldoses and Ketoses

Aldoses contain an aldehyde functional group at carbon-1 (C-1), whereas ketoses contain a carbonyl group that is almost always located at carbon-2 (C-2). C-1 for aldoses and C-2 for ketoses are the reactive centers for these molecules and are known as the **anomeric carbon atoms**. Figure 1.1 shows the structure for D-glucose, D-fructose, and, in addition, D-arabinose. Sugars have common or trivial names with historical origins from chemistry, medicine, and industry. There is also a systematic procedure for naming sugars (some examples are shown in Table 1.1). Glucose is also commonly known as dextrose. In systematic nomenclature, its suffix is hexose, indicating a 6-carbon aldose sugar, and the prefix is *gluco-*, which shows the orientation of the hydroxyl groups around carbons 2–5. The symbol D refers to the orientation of the hydroxyl group on C-5, the

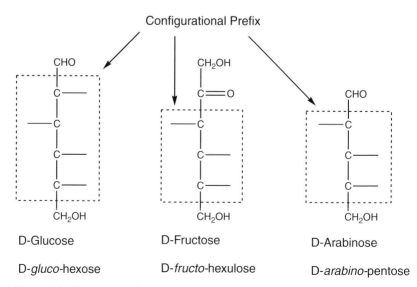

Figure 1.1   Structure and nomenclature of glucose, fructose, and arabinose.

Table 1.1   Trivial and Systematic Names of Selected Sugars

Trivial (or Common)	Systematic[a]
D-Erythrose	D-*erythro*-tetrose
D-Threose	D-*threo*-tetrose
D-Arabinose	D-*arabino*-pentose
D-Lyxose	D-*lyxo*-pentose
D-Ribose	D-*ribo*-pentose
D-Xylose	D-*xylo*-pentose
D-Allose	D-*allo*-hexose
D-Altrose	D-*altro*-hexose
D-Galactose	D-*galacto*-hexose
D-Glucose	D-*gluco*-hexose
D-Gulose	D-*gulo*-hexose
D-Idose	D-*ido*-hexose
D-Mannose	D-*manno*-hexose
D-Talose	D-*talo*-hexose

[a]In the systematic name, the configurational prefix is italicized, and the stem name indicates the number of carbon atoms in the molecule.

highest numbered asymmetric carbon atom, also known as the **reference carbon atom**. Since fructose (also known as levulose) has just three asymmetric carbon atoms, its configurational prefix is the same as that for the pentose sugar arabinose. Thus, the systematic name for glucose is D-*gluco*-hexose and fructose is D-*arabino*-hexulose.

Configurations of Aldose Sugars

Figure 1.2 shows all possible configurations around the asymmetric carbon atoms for the triose, tetrose, pentose, and hexose D-aldose sugars. **Diastereoisomers** are molecular isomers that differ in configuration about one or more asymmetric carbon atoms; there are eight hexose diastereoisomers. **Epimer** is yet another term in sugar chemistry that refers to diastereoisomers that differ in configuration about only one asymmetric carbon atom (e.g., D-galactose is the 4-epimer of D-glucose). The term has historical significance because the melting point of dinitrophenylhydrazone derivatives was a classical procedure used in identifying sugars. The 2-epimers (e.g., glucose and mannose, allose and altrose, etc.) gave identical dinitrophenylhydrazone derivatives.

D- vs. L-Sugars

L-sugars are the mirror images of D-sugars. Figure 1.3 depicts the structures of D- and L-glucose in the Fischer and conformational projections. (Note: When drawing an L-sugar, the orientation of the hydroxyl groups

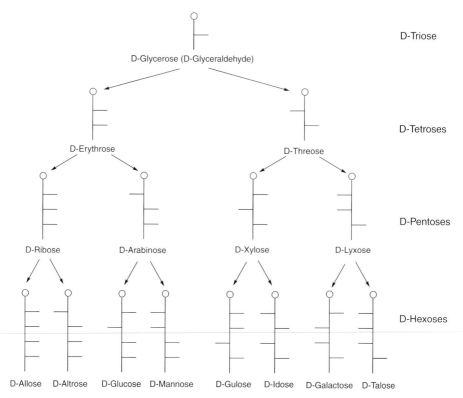

Figure 1.2   **Structures of the D-aldoses containing from 3 to 6 carbon atoms as depicted by the Rosanoff shorthand convention.**

on every asymmetric carbon atom need to be reversed; a frequent error is to only change the orientation on the reference carbon atom.) D- and L-glucose are **enantiomers**, nonsuperimposable stereoisosmeric molecules that are mirror images. L-sugars occur rarely in nature. A pair of

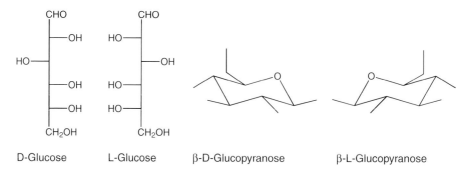

Figure 1.3   **D- and L-glucose as depicted in the Fischer and conformational projections.**

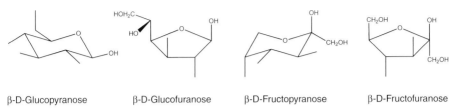

β-D-Glucopyranose   β-D-Glucofuranose   β-D-Fructopyranose   β-D-Fructofuranose

Figure 1.4   Furanose and pyranose structures of β-D-glucose and β-D-fructose.

enantiomers are identical in chemical reactivity, and they have the same taste properties. They are handled differently in biological systems, however. Although humans absorb L-sugars, L-sugars are not metabolized and thus have no caloric value.

Different Ways of Depicting Sugar Structures

Fischer, Haworth, Mills, and Conformational Structures

It is important to realize that pentose and hexose sugars exist as ring forms, and only a small amount will occur in the acyclic form. Thus, the Fischer projections shown in Figure 1.2 are in no way representative of actual structural form. The Fischer projections are useful in that they provide a code for the orientation of hydrogen and hydroxyl substituents on adjacent carbon atoms. Sugars exist as both six-membered (pyranose) and five-membered (furanose) ring structures (Figure 1.4). The furanose ring is relatively planar, whereas the pyranose ring is "puckered," existing in chair and boat forms. Figure 1.5 illustrates β-D-glucopyranose by the Haworth, Mills, and conformational conventions. The Haworth representation is troublesome because it shows the pyranose ring as being planar. The Mills structure takes an aerial view with hydroxyl groups depicted as

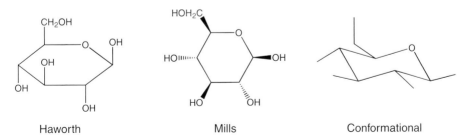

Haworth   Mills   Conformational

Figure 1.5   β-D-glucopyranose as depicted by the Haworth, Mills, and conformational conventions.

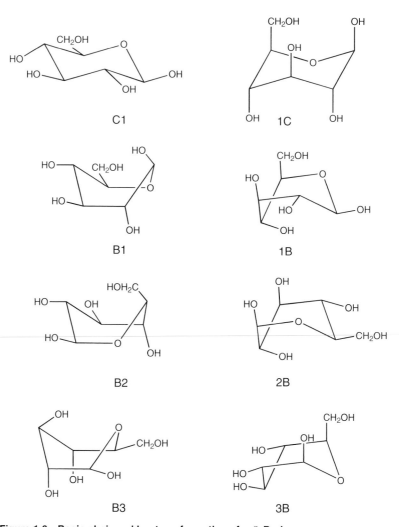

Figure 1.6   Basic chair and boat conformations for β-D-glucopyranose.

being either "upward" or "downward." The conformational representation shows perspective where the ring oxygen is remote, equatorial substitutents are drawn at an angle, and axial substituents are vertical. The conformational representation is highly preferred and will be used predominately throughout this book.

There are two stable chair forms and six stable boat forms for pyranose sugars; the forms for β-D-glucopyranose are shown and named in Figure 1.6. Sugar conformation is best visualized and demonstrated using molecular models. Relatively inexpensive and stereochemically accurate kits are available, and practice with them is highly recommended. Showing three-dimensional structure on the two-dimensional printed page has obvious limitations. Given the conformational structure of a pyranose sugar, one should be able to make the correct molecular model; and, vice versa, given a molecular model, one should be able to draw the correct conformational structure. An exercise is included in the Appendix for making molecular models of sugar molecules with molecular model kits.

Classifying Sugars by Compound Class—Hemiacetals, Hemiketals, Acetals, and Ketals

Aldehydes and ketones will condense with alcohols to form hemiacetals and hemiketals. In excess alcohol and dehydrating conditions, further condensation will take place to form acetals and ketals. A generalized reaction of an aldehyde with alcohol is shown in Figure 1.7, and the generalized structures of hemiacetals, hemiketals, acetals, and ketals are shown below.

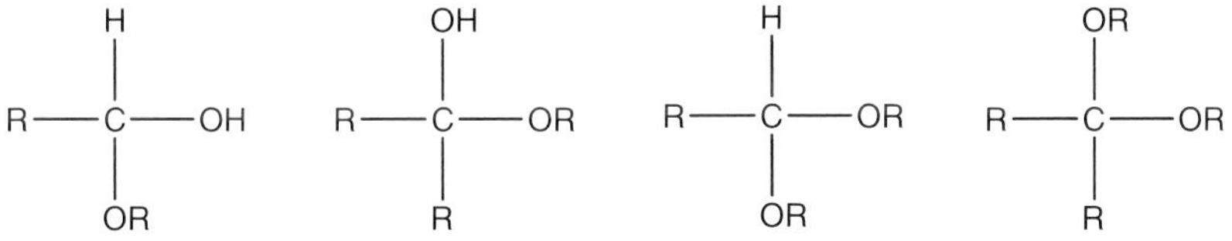

Structure 1.1   Hemiacetal, Hemiketal, Acetal, Ketal

Pyranose and furanose ring forms of sugars are hemiacetals or hemiketals and are formed from intramolecular condensation of the carbonyl group with a hydroxyl substituent. When this reaction occurs, two products are formed, as the hydroxyl group on the anomeric carbon has two possible orientations, as illustrated by α- and β-D-glucopyranose. They have different chemical properties, with the β-form being more stable because of its equatorial disposition. They are known as **anomers**, diastereoisomers that differ in configuration only at C-1 for aldoses and C-2 for ketoses.

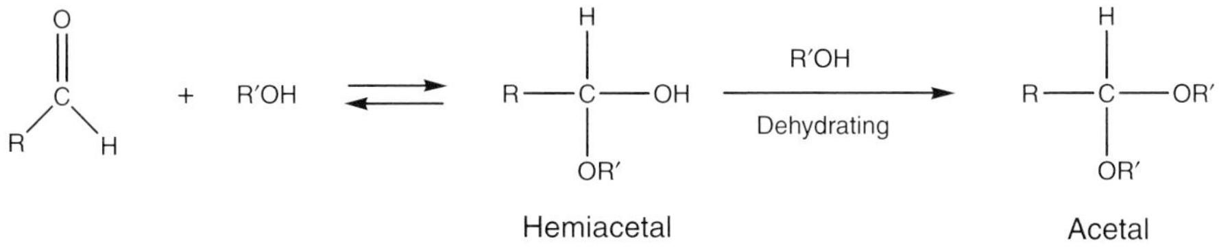

Figure 1.7   Reaction of an aldehyde with an alcohol to form a hemiacetal and an acetal.

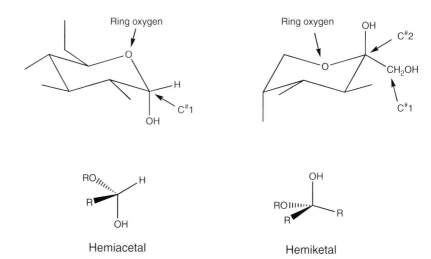

Figure 1.8   Location of the hemiacetal and hemiketal functional groups in β-D-glucopyranose and β-D-fructopyranose.

Being able to recognize hemiacetal, hemiketal, acetal, and ketal functional groups is critical to carbohydrate chemistry. The chemical reactivity and functionality of different sugars is directly related to the presence of those functional groups. Figure 1.8 illustrates the location of the hemiacetal and hemiketal functional groups for β-D-glucopyranose and β-D-fructopyranose. To correctly recognize these functional groups, one should proceed as follows:

- Locate the anomeric carbon atom. This will be C-1 for aldose sugars and C-2 for ketose sugars.
- The anomeric carbon atom will be to one side of the ring oxygen.
- The carbon on the other side of the ring oxygen may or may not be the reference carbon atom. The reference carbon atom is C-5 for hexoses and C-4 for pentose sugars. Once the anomeric carbon atom has been located, it helps to number all carbon atoms.

An exercise for identifying hemiacetal, hemiketal, acetal, and ketal functional groups in sugars is included in the Appendix.

Structure and Nomenclature of Disacchaarides

Disaccharides are either reducing or nonreducing, and a reducing disaccharide can be identified by the presence of a hemiacetal or hemiketal functional group. Disaccharides are formed from the reaction of an anomeric carbon atom with the hydroxyl group of another sugar. This acetal or ketal linkage is also referred to as a glycosidic linkage. The hydroxyl group reacting with the anomeric carbon may be an alcohol functional group of another sugar, or it could be the hydroxyl substituent located on the anomeric carbon. When the condensation is between two anomeric carbons, the compound will not contain a hemiacetal or hemiketal functional group, and it will be a nonreducing sugar. Reducing disaccharides are systematically named by having the nonreducing sugar moiety be a substituting group on the reducing sugar. The nature of the glycosidic linkage, whether α or β, the number of the carbon atom where the sugar is substituted, and the ring size all need to be indicated. Importantly, reducing disaccharides have an -ose suffix. For example, the systematic name for lactose is 4-O-β-D-galactopyranosyl-D-glucopyranose. Reducing disaccharides will have both α and β forms, the designation being for the orientation of the anomeric hydroxyl, not the glycosidic linkage. If the glycosidic linkage is changed from β to α, a different sugar is formed (e.g., the sugar will no longer be lactose). The structures along with systematic and trivial names for disaccharides that are important in foods are shown in Figure 1.9. Lactose is also known as milk sugar and has nutritional relevance because of lactose intolerance (inability to digest lactose) that is prevalent in some human populations.

Sucrose is a nonreducing sugar; hence, both anomeric carbons participate in the glycosidic linkage. There is no need to indicate which carbons are involved in systematic nomenclature, because it is known that both C-1 of glucose and C-2 of fructose are involved. Either sugar can be the substituting sugar, and the nature of both glycosidic linkages needs to be given. The suffix for nonreducing sugars is -ide. The correct name for sucrose is either α-D-glucopyranosyl-β-D-fructofuranoside or β-D-fructofuranosyl-α-D-glucopyranoside. The relevance of sucrose to food technology, nutrition, and world trade is monumental. Maltose (4-O-α-D-glucopyranosyl-D-glucopyranose) is the building unit of starch, whereas cellobiose (4-O-β-D-glucopyranosyl-D-glucopyranose) is the building unit for cellulose. Trehalose (α-D-glucopyranosyl-α-D-glucopyranoside) consists of two glucose molecules linked head-to-head and is nonreducing. It occurs in mushrooms and insects and is metabolized by humans. (This gives credence to the hypothesis that insects played an important role in man's diet from an evolutionary perspective.) The nonreducing and moderate sweetness properties of trehalose make possible some innovative food applications. The structures and names for an extensive list of known disaccharides are given in Unit 9 of the Appendix.

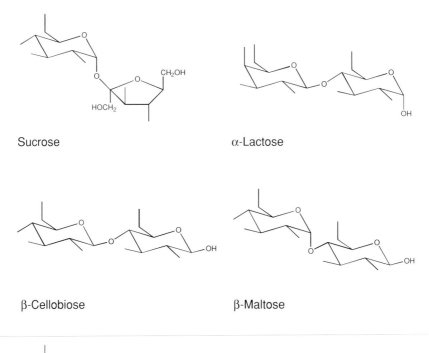

Sucrose   α-Lactose

β-Cellobiose   β-Maltose

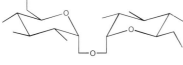

Trehalose

Figure 1.9   Structures of five disaccharides of particular importance in foods: sucrose (β-D-fructofuranosyl-α-D-glucopyranoside   or   α-D-glucopyranosyl-β-D-fructofuranoside), lactose (4-O-β-D-galactopyranosyl-D-glucopyranose), maltose (4-O-α-D-glucopyranosyl-D-glucopyranose), cellobiose (4-O-β-D-glucopyranosyl-D-glucopyranose), and trehalose (α-D-glucopyranosyl-α-D-glucopyranoside).

Structure and Optical Activity

Enantiomers were previously defined as nonsuperimposable stereoisomeric molecules that are mirror images. This is best demonstrated with molecular models by making a model of β-D-glucopyranose, for example, and then also creating its mirror image, β-L-glucopyranose. The two molecules can be visualized as mirror images by placing them head-to-head or, alternatively, placing one below the other. They are nonsuperimposable, meaning that they cannot be stacked with C-1 being over C-1,

Table 1.2   Specific Rotation $[\alpha]_D^{20}$ of Selected Pairs of D- and L-Sugars at Equilibrium

Sugar	$[\alpha]_D^{20}$	Sugar	$[\alpha]_D^{20}$
D-Glyceraldehyde[a]	+13.5°	L-Glyceraldehyde	+13.5°
D-Threose[b]	−12°	L-Threose	+12°
D-Ribose[c]	−21°	L-Ribose	+21°
D-Arabinose[c]	−105°	L-Arabinose	+105°
D-Xylose[c]	+19°	L-Xylose	−19°
D-Lyxose[c]	−14°	L-Lyxose	+14°
D-Allose[d]	+14°	L-Allose	−14°
D-Glucose[c]	+52°	L-Glucose	−52°
D-Mannose[c]	+14°	L-Mannose	−14°
D-Gulose[e]	−26°	L-Gulose	+26°
D-Galactose[c]	+80°	L-Galactose	−80°
D-Talose[c]	+21°	L-Talose	−21°
D-Fructose[c]	−92°	L-Fructose	+92°

Sources: [a]Weast 1970; [b]Hockett 1935; [c]Pigman and Isbell 1968; [d]Bates 1942; [e]Isbell 1937.

C-2 over C-2, etc. They are asymmetric, they have "handedness," and they are **chiral**. They have the property of being able to rotate plane-polarized light in equal amounts but in opposite directions. (**Optical activity** is the ability to rotate plane-polarized light.) The specific rotations of several D- and L-sugars are shown in Table 1.2. Note that some D-sugars (e.g., D-glucose and D-xylose) are **dextrorotary**, having the ability to rotate plane-polarized light to the "right" or in a clockwise direction, whereas other D-sugars (e.g., D-arabinose, D-ribose, D-gulose, and D-fructose) are **levorotary**, having the property of rotating plane-polarized light to the "left" or in a counter-clockwise direction. Thus, being a D- or L-sugar does not predict whether a sugar is dextrorotary or levorotary. The designation D and L should not be confused with *d* and *l*, which denotes having optical activity that is dextrorotary or levorotary. (The trivial names dextrose and levulose for glucose and fructose, respectively, have their origin from D-glucose being dextrorotary and D-fructose being levorotary.) The appendix contains instructions for demonstrating the existence of polarized light and the ability for sugar solutions to rotate it using plastic polarizing material, a light source such as a flashlight, and sugar solutions.

Anecdotal information

D- and L-glucose have identical physical and chemical properties (i.e., the same solubility, reactivity, and taste properties). L-Sugars are not used by humans, however, and are noncaloric. In 1981, Biospherics Inc. of

Beltsville, MD received a patent for use of several L-sugars as low-calorie sweeteners in foods (Levin and Zehner 1991). Their petition to the FDA for its use, however, was denied. The medical and nutritional communities supported the FDA's action, as the wisdom of having considerable quantities of reactive L-sugars circulating in the bloodstream from consumption of L-sugars was questionable.

Specific rotation is a physical property of sugars that is useful in identifying sugars and measuring their concentration. Measurements are taken with a polarimeter that has a source of plane-polarized light, and a prism that enables measurement of the degree that a sugar solution in a cell through which the light is being transmitted will be rotated. The equation for calculating specific rotation is shown below. α is the observed rotation in degrees, c is the concentration of the sugar solution in g/mL, l is the path length of the cell in dm, and D is the wavelength for the D line of Na, 589.3 nm.

$$\text{Specfic Rotation} = [\alpha]_D = \alpha/c \times 1 \tag{1.2}$$

Table 1.3 gives the specific rotation for several different sugars. Not only do different sugars (e.g., sucrose vs. trehalose) have different specific

Table 1.3   Specific Rotation $[\alpha]_D^{20}$ of Selected Sugars[a]

Sugar	$[\alpha]_D^{20}$ (Initial)	$[\alpha]_D^{20}$ (Final)
α-D-Glucose	+112°	+52°
β-D-Glucose	+19°	+52°
α-D-Galactose	+144°	+80°
β-D-Galactose	+52°	+80°
α-D-Mannose	+34°	+15°
β-D-Mannose	−17°	+15°
β-D-Fructopyranose	−133°	−92°
β-D-Fructofuranose	−21°	−92°
α-D-Xylose	+92°	+19°
β-D-Xylose	−20°	−19°
α-D-Lactose	+90°	+55°
β-D-Lactose	+35°	+55°
α,α-Trehalose	+197°	+197°
Sucrose	+66.5°	+66.5°
Raffinose	+124°	+124°
Stachyose	+148°	+148°
Invert Sugar		−20°
α-L-Glucose	−96°	−52°
Levoglucosan	−67°	−67°

[a]Source: Shallenberger and Birch 1975; Shallenberger 1982.

rotation, but different anomeric and ring forms of the same sugar also have different specific rotation. When crystalline α-D-glucopyranose is dissolved in water, its initial rotation will approach +112. Reducing sugars in aqueous solution will "mutarotate" until they come to an equilibrium mixture of α and β, furanose and pyranose, and open-chain forms. Nonreducing sugars (e.g., sucrose and trehalose) will not mutarotate, that is their initial and final rotations will be the same because they are not undergoing structural transformations. Observing whether a crystalline sugar will mutarotate is another method for determining whether it is reducing or nonreducing. The Appendix includes a laboratory exercise on polarimetry of sugars.

Whether a sugar is D or L will not predict whether it is dextrorotary or levorotary. Being *d* or *l* is a fairly reliable predictor of sugar conformation, however. D-Pyranose sugars that favor the C-1 conformation are dextrorotary, whereas those favoring the 1-C conformation are levorotary.

A Systematic Procedure for Determining Conformation (C-1 or 1-C), Chiral Family (D or L), and Anomeric Form (α or β) of Sugar Pyranoid Ring Structures

The conformational structures of D-glucopyranose and D-fructopyranose are typically shown as depicted in Figures 1.3, 1.4, and 1.8. Conventionally, the structures are drawn with C-1 oriented to the right of the ring oxygen for glucose and C-2 to the right of the ring oxygen of fructose. The molecules, of course, will not always be oriented in this direction, and drawing them with a different spatial orientation, as per the structures of glucose in trehalose (Figure 1.9), can be challenging. Drawing L-sugars can be particularly troublesome. One approach is to make the molecular model and then draw the conformational structure. Shallenberger and others (1981) developed a systematic procedure for determining the conformation, chiral family, and anomeric form for sugar pyranoid ring structures. The step-wise procedure requires that one locate the anomeric carbon atom and the reference carbon atom. One determines whether the structure is drawn with the ring oxygen above or below the plane of the ring and whether the structure is oriented clockwise or counterclockwise. Plus (+) and minus (−) assignments are made for these different features, and a series of algebraic multiplications will determine whether the ring structure is C-1 or 1-C, D or L, and α or β. The C-1 conformation of β-D-glucopyranose serves as the basic reference with all assignments being positive. The Appendix contains an exercise for determining the above features for selected sugars. The protocol is useful for determining whether or not a structure is properly drawn, and the process requires correct identification of anomeric and reference carbon atoms.

Structure and Nomenclature of Sugar Derivatives with Relevance to Food Chemistry

Glycols (Alditols)

Glycols (**alditols**) are sugar alcohols, the carbonyl group being reduced to an alcohol. The rules for systematic nomenclature are to replace the -ose suffix with –itol; that is, the alditol derivative of D-glucose is D-glucitol. The trivial name for D-glucitol is sorbitol, which is widely distributed in nature. Sorbitol is present in trace amounts in many fruits and in substantial quantities in apples (5% of total sugars), pears (18%), cherries (15–20% of total sugars), and chokeberries (*Aronia melanocarpa*; up to 50% of total sugars). Commercially, sorbitol is manufactured by catalytic reduction of glucose (Figure 1.10). The body will passively absorb sorbitol, and it is converted to glucose in the liver, thus it is **glucogenic** (the body can convert it to glucose) and caloric. Some diabetics can accommodate sorbitol because of its slow absorption rate; it is found as an ingredient in diabetic confections and jams. Sorbitol that is not absorbed in the small intestine will go to the colon, where it can cause diarrhea because of its **hygroscopic** (affinity for water) quality. Sorbitol is used to retard sugar crystallization in confections, and to provide viscosity and body in some beverage formulations. Because it is nonreducing, it will not react with amino acids in Maillard browning, and for that reason, one application is to use it as a

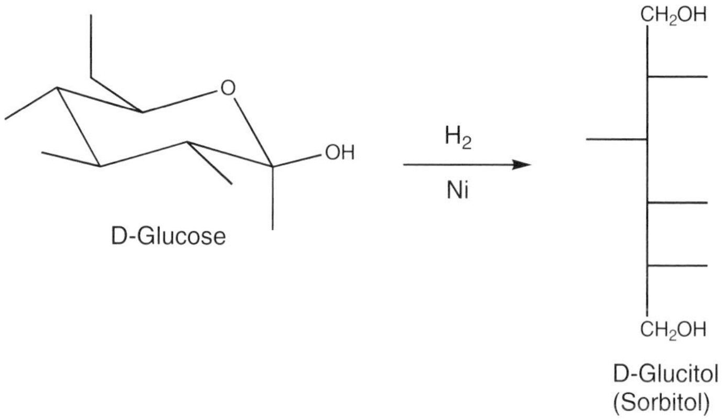

Figure 1.10   Catalytic reduction of glucose to form sorbitol.

cryoprotectant in surimi manufacture, in which a white appearance is a critical quality factor.

Xylitol has some unique properties and is approved for use in some special dietary foods.

$$
\begin{array}{c}
CH_2OH \\
| \\
C\text{——} \\
| \\
\text{——}C \\
| \\
C\text{——} \\
| \\
CH_2OH
\end{array}
$$

Structure 1.3   Xylitol

It is **noncariogenic** (will not support the growth of bacteria, such as *Streptococcus mutans*, in dental plaque). Food applications are in chewing gums and some candies. It has a negative heat of solution, its endothermic enthalpy of dissolution being approximately ten times that of sucrose. Thus, it will provide a cooling effect when dissolved in the mouth. Xylitol has symmetry and is not optically active. Therefore, it is neither D nor L with respect to its configuration. Additional alditols that are **achiral** (without handedness) and optically inactive are glycerol, erythritol, ribitol, allitol, and galactitol.

Glyconic, Glycuronic, and Glycaric Acids

Mild oxidation of an aldose sugar will produce a **glyconic acid** that can cyclize intramolecularly to form lactones. Aldonic acids are named by replacing the suffix -ose with -onic acid. Figure 1.11 shows oxidation of glucose to form D-gluconic acid, which subsequently reacts intramolecularly to form δ-gluconolactone and γ-gluconolactone. δ-Gluconolactone is an approved food additive and has the property of being neutral but becomes acid when dissolved in water. This slow release of acid finds application in certain leavening systems. Salts of gluconic acid are water-soluble; calcium gluconate provides a means of increasing water solubility to effectively deliver calcium in dietary calcium supplements. Ferrous gluconate is an approved color additive in the United States; its major application is for black olives. The gluconate function renders water solubility to the salt, which permits iron to be distributed throughout the fruit where it can complex with phenolics to produce a black color.

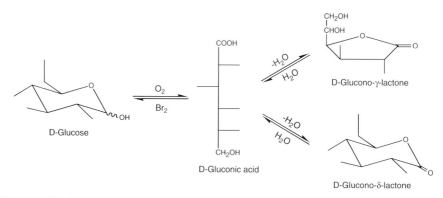

Figure 1.11   Oxidation of D-glucose to form D-gluconic acid, followed by intramolecular cyclization to form γ-gluconolactone and δ-gluconolactone.

Oxidation of the primary alcohol group of an aldose sugar to a carboxylic acid produces a **glycuronic acid**. Glycuronic acids are named by replacing the suffix -ose of an aldose sugar with -uronic acid. The structure of D-galacturonic acid, a major constituent of pectin, is shown below.

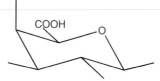

Structure 1.4   D-Galacturonic acid

Oxidation of both the aldehyde and primary alcohol functional groups of an aldose sugar to carboxylic acids gives a sugar derivative known as an **aric acid**. Galactaric acid (also known as mucic acid) serves as an example.

$$
\begin{array}{c}
\text{COOH} \\
|\\
\text{C}\!-\!-\!- \\
|\\
-\!-\!-\text{C} \\
|\\
-\!-\!-\text{C} \\
|\\
\text{C}\!-\!-\!- \\
|\\
\text{COOH}
\end{array}
$$

Structure 1.5   Galactaric acid

Galactaric acid can form mono- or dilactones, and it is used in some baking powders as a slow-release acidulant. Galactaric acid has symmetry and is not optically active. Additional aric acids will have this property, as discussed previously for alditol derivatives.

Deoxy Sugars

Substitution of a hydroxyl group with a proton forms a doxy sugar, which is named by indicating the carbon number containing no oxygen as a prefix, for example, 2-deoxy-D-ribose (an important constituent of nucleic acids) and 6-deoxy-L-mannose (trivial name = rhamnose). Rhamnose has an important structural role in the polysaccharide pectin and is frequently found as a phenolic glycoside in plants.

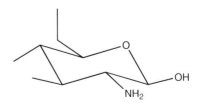

Structure 1.6   Rhamnose (6-deoxy-L-mannose)

Amino Sugars and Glycosyl Amines

An **amino sugar** is a sugar derivative that has the hydroxyl group on any carbon other than the anomeric carbon atom replaced with an amino group. They are named as a substituted deoxy sugar, placing in alphabetical order the substituting group and the prefix deoxy (e.g., 2-amino-2-deoxy-D-glucose).

Structure 1.7   D-glucosamine (2-amino-2-deoxy-D-glucose)

Chitin, a polymer of D-glucosamine, is a major structural component of the exoskeleton of crustacea. Better use of the wastes from crab, lobster, and shrimp processing is a continuing research topic for seafood processors. Chitin is also present in insect exoskeleton and in molds. Its detection has been proposed as an index for mold contamination in foods (Cousin 1996).

Glycosyl amines are sugar derivatives that have the anomeric hydroxyl group replaced with an amino function. They are named as a substituted amine, for example, β-D-glucosyl amine and β-D-glucosyl glycine. Not only are amino sugars and glycosyl amine named differently, they are also very different with respect to chemical reactivity.

Structure 1.8   β-D-glucosyl amine, β-D-glucosyl glycine

Glycosides

Glycosides are sugar derivatives that have been formed by reaction of a reducing sugar with an alcohol or phenol, resulting in substitution of the anomeric hydroxyl with an OR group of the alcohol or phenol. They are named by first listing the substituting group followed by the sugar with the suffix -ose replaced with -ide. Two examples are methyl-β-D-glucopyranoside and the anthocyanin pigment cyanidin-3-β-D-glucoside.

Structure 1.9   Methyl-β-D-glucopyranoside and cyanidin-3-β-D-glucoside

Sugar Ethers and Sugar Esters

Sugar ethers are derivatives where the hydroxyl group on carbon atoms other than the anomeric carbon atom is substituted with an -OR group. They are named by designating the number of the substituted carbon, followed by -O- and the substituting group, and then the sugar. If the sugar is a reducing sugar, the -ose suffix is retained. For example, 4-O-methyl-β-D-glucoopyranose. Trimethylsilyl ethers are sugar derivatives that permit sugar analysis by gas-liquid chromatography (see Chapter 6).

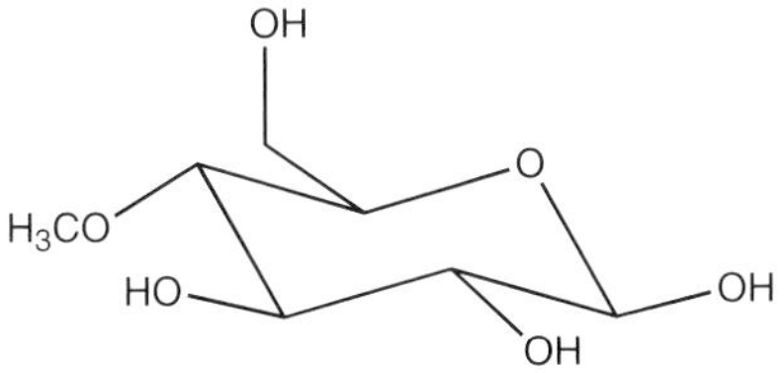

Structure 1.10   4-O-methyl-β-D-glucopyranose

Trimethylsilyl - 2,3,4,6 - O - tetra - trimethylsilyl - β - D - glucopyranoside is shown below. Note that the trimethylsilyl function on the anomeric carbon is a glycoside, and so named.

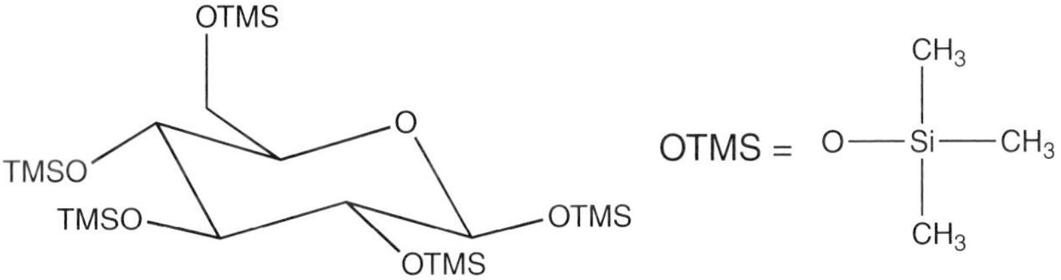

Structure 1.11   Trimethylsilyl-2,3,4,6-O-tetra-trimethylsilyl-β-D-glucopyranoside

All hydroxyl groups of sugars must be substituted with trimethylsilyl groups in order for the sugar to have sufficient volatility for gas-liquid chromatographic analysis.

Sugar hydroxyls can combine with organic and inorganic acids to form **sugar esters**. Sugar acetates and succinates occur in nature, and sugar phosphates are important metabolic intermediates. Sucrose polyester (trade name Olestra^{TM}) is an approved food additive in which sucrose is fully esterified with fatty acids. It is a lipophilic, nondigestible and nonabsorbable fat-like molecule. Hence, it is noncaloric and approved for limited use in foods as a fat substitute.

R = long-chain fatty acid

Structure 1.12   Sucrose polyester (trade name Olestra™)

Vocabulary

Achiral—Without handedness

Anomers—Diastereoisomers differing in configuration only at C-1 for aldoses and C-2 for ketoses

Anomeric carbon atom—C-1 for aldoses and C-2 for ketoses

Chiral—Possessing handedness, having no element of symmetry; being asymmetric

Configuration—The arrangement in space of substituents about the tetrahedral carbon atom

Conformation—The shape of a molecule that may be altered simply by rotation (twisting) about single covalent bonds

D-Sugars—Sugars in which the hydroxyl group on the highest numbered asymmetric carbon atom is oriented to the "right"

Dextrorotary—Having the ability to rotate plane-polarized light to the "right" or in a clockwise direction

Diastereoisomers—Isomers that differ in configuration about one or more asymmetric carbon atoms (e.g., glucose and galactose)

Epimers—Diastereoisomers that differ in configuration about only one asymmetric carbon atom

Glucogenic—The ability to be metabolically converted to glucose

Glycose—Generic name for sugars

L-Sugars—Sugars in which the hydroxyl group on the highest numbered asymmetric carbon atom is oriented to the "left"

Levorotary—Having the ability to rotate plane-polarized light to the "left" or in a counterclockwise direction

Optical activity–The ability to rotate plane-polarized light

Reference carbon atom—The highest numbered asymmetric carbon atom (e.g., C-5 for hexoses and C-4 for pentose sugars)

References

Bates FJ. 1942. *Polarimetry, saccharimetry and the sugars*. Washington, D.C.: U.S. Government Printing Office. [*Note: Perusal of this book should generate admiration for monumental work in sugar chemistry by the U.S. National Bureau of Standards.*]

Cousin MA. 1996. Chitin as a measure of mold contamination of agricultural commodities and foods. *J Food Protect* 59:73–81.

Hockett RC. 1935. The chemistry of tetrose sugars. II. The degradation of *d*-Xylose by the method of Wohl. The rotation of *d*-threose. *J Am Chem Soc* 57:2265–2268.

Isbell HS. 1937. Configuration of the pyranoses in relation to their properties and nomenclature. *Bur Stand J Res* 18:505–541.

Levin GV, Zehner LR. 1991. L-Sugars: Lev-O-Cal. In: Nabors LO, Gelardi RC, editors. *Alternative sweeteners, 2nd ed.* New York: Marcel Dekker, pp. 117–125.

Pigman W, Isbell HS. 1968. Mutarotation of sugars in solution. In: Wolfrom ML, Tipson, RS, editors. *Advances in carbohydrate chemistry.* New York: Academic Press, pp. 11–57.

Shallenberger RS. 1982. *Advanced sugar chemistry*. Westport, CT: AVI Publishing.

Shallenberger RS, Birch GG. 1975. *Sugar chemistry*. Westport, CT: AVI Publishing.

Shallenberger RS, Wrolstad RE, Kerschner LE. 1981. Calculation and specification of the multiple chirality displayed by sugar pyranoid ring structures. *J Chem Educ* 58(8):599–601.

Weast RC. 1971. *Handbook of chemistry and physics, 51st edition*. Cleveland, OH: The Chemical Rubber Co.

2 Sugar Composition of Foods

Introduction

There is heightened interest and confusion amongst consumers, athletes, health professionals, nutritionists, and food technologists about sugars in foods. This is not surprising considering the roles that dietary sugars may play in diabetes, obesity, lactose intolerance, dental caries, and athletic performance. Caretakers find they need to adjust their cooking and food purchase habits when family members are diagnosed with diabetes and lactose intolerance. Food technologists find they need to adjust product formulations to provide healthier products in light of new nutritional knowledge and to cater to consumer demands and niche markets. There is no shortage of information and misinformation about dietary sugars from

Food Carbohydrate Chemistry, First Edition. Ronald E. Wrolstad.
© 2012 John Wiley & Sons, Inc. Published 2012 by John Wiley & Sons, Inc.

the Internet, television talk shows, and the popular press. It is prudent that food professionals have a working general knowledge of sugar composition of foods and where to access accurate compositional data.

Sugar Content of Foods

The USDA Nutrient Data Laboratory recently analyzed approximately 850 food products for sugar content (USDA 2009). Using high-performance liquid chromatography (HPLC) they measured the concentrations of the following individual sugars: glucose, fructose, galactose, sucrose, lactose, and maltose, and reported the summation as sugar content. Table 2.1 presents the sugar content of selected food items from the USDA database, organizing the foods according to high (37.5–100 g sugar/100 g food), medium (10.5–18.6 g sugar/100 g food), and low (0–6.5 g/100 g food).

The category of high-sugar foods includes sugar itself, candies, dried fruits, jams, and preserves. The high sugar content of dried fruit, jams, and preserves is brought about by water removal and also by sugar addition in the case of jams and preserves. They are historical examples of food preservation by reducing the water activity so these products will not support microbial growth. Sugar-sweetened cereals that appeal to children are in the high category, which has concerned many parents and health professionals. Most breakfast cereals are in the medium category; however, people will always have the option of adding table sugar ad libitum. Other foods in the medium category include most fruits, fruit juices, ice creams, and soft drinks. Vegetables, nuts, milk, many snack foods, wine, and beer have low sugar content.

Composition of Sweeteners

Cane and Beet Sugar

Sugarcane, genus *Saccharum*, is native to South Asia and Southeast Asia, where it was first cultivated. Arabs in the eighth century introduced it to the Mediterranean countries. It was among the early crops brought to the Caribbean Islands and the southern United States by the Spanish. A combination of climate, slave labor, and industrial technology set the stage for sugar to become a major trade commodity and common food staple. In the refining process, shredded sugarcane is mixed with water and pressed (Pancoast and Junk 1980). The pH of the cane juice is adjusted to 7 with lime to prevent sucrose hydrolysis. Clarified juice

Table 2.1   Sugar Content[a] of Selected Foods

High-Sugar Foods (37.5–100 g sugar/100 g food)	
Granulated sugar	100
Brown sugar	96.3
Honey	82.1
Candies (fudge, jelly beans, M&M, Mars, gumdrops)	59.0–73.8
Marshmallows	57.6
Dates	63.3
Raisins	59.2
Jams and preserves	48.5
Sugar-sweetened cereals (e.g., Kellogg's Frosted Flakes)	38.0
Sugar cookies	37.7
Medium-Sugar Foods (10.5–18.6 g sugar/100 g food)	
Chocolate pudding	17.8
Milk shakes	17.8
Soft-serve ice milk	16.4
Cereals (e.g., General Mills whole-grain Total)	15.8
Grapes (Thompson seedless)	15.5
Cereals (e.g., General Mills Wheaties)	14.0
Bananas	12.2
Bread (whole, mixed-grain)	10.9
Apple juice	10.9
Carbonated soft drinks	10.6
Low-Sugar Foods (0–6.5 g sugar/100 g food)	
Canned baked beans	6.5
Watermelon	6.2
Sweet potato	5.7
Nonfat milk	5.1
Reduced fat milk (2%)	5.1
Cashews (dry roasted)	5.0
Carrots (raw)	4.8
Almonds	4.8
White bread	4.7
Yogurt (plain)	4.7
Peanuts	4.2
Potato chips	4.1
Tomatoes	2.6
Avocados	2.4
Mushrooms	2.2
Mashed potatoes	1.5
White table wine	0.98
Red table wine	0.62
Beer (light)	0.09
Beer	0

[a]Sum of glucose, galactose, fructose, sucrose, lactose, and maltose as determined by HPLC.
Source: USDA 2009.

is concentrated in a multiple effect evaporator to syrup, 60% by weight. The syrup is further concentrated and seeded with crystalline sucrose. Sugar crystals are separated from molasses by centrifugation. Raw sugar is further refined through a series of washing, concentration, salt removal, and filtration steps. The purified syrup is concentrated to supersaturation and repeatedly recrystallized under vacuum.

Sugar beets (*Beta vulgaris* L.), which are grown in temperate climates, are the other major source for sugar. Selective breeding at the beginning of the nineteenth century produced a beet that contained 5–6% sucrose by weight, which was sufficient to stimulate establishment of the first beet sugar refinery in 1801. Today's commercial varieties contain a dry mass of 15–20% sucrose by weight. In beet sugar refining, the beets are sliced into thin strips and subjected to counter-current extraction with hot water. The pH of the raw juice is adjusted with lime, and the subsequent concentration, purification, and crystallization steps are similar to that for sugar cane.

Types of Sugar Products Available

Granulated sugar—Pure sugar is available in various crystal or granular sizes (e.g., coarse, medium, fine, extra fine, superfine, etc).

Brown sugar—Historically, brown sugar was less-refined sugar crystals that contained entrapped molasses. Today, most commercial brown sugar is produced by adding various amounts of molasses to white sugar to give varying shades of color from light yellow to dark brown. They contain varyious amounts of amorphous sugar, ash, and water.

Powdered sugar—Confection (powdered) sugar is finely ground sucrose that contains 3% cornstarch as an anticaking ingredient.

Liquid sugar—Aqueous solutions of refined sucrose are available for some industrial applications.

Invert sugar—Sucrose will readily undergo enzymic or acid hydrolysis to produce glucose and fructose. It is commercially available in liquid forms of 50% glucose and 50% fructose. It is also available with the approximate composition of 50% sucrose, 25% glucose, and 25% fructose.

Honey

Honey has been a popular sweetener since biblical times. Fructose (38%) and glucose (31%) are the major sugars, but it also contains sucrose, maltose, and other oligosaccharides. Its water content is 17%, and it contains 0.02–1% minerals. The floral nectar source contributes heavily to its flavor. There are USDA grades of A, B, C, and substandard for honey based on a minimum percent soluble solids, flavor and aroma, and absence of defects (USDA 1985). In addition, there are official USDA color standards ranging from "water white" to "dark amber." The viscosity and hygroscopicity of honey make for handling frustrations in commercial product formulation. Dried free-flowing honey granules and products are available commercially that have been manufactured by spray drying and combinations of microwave vacuum and freeze-drying. They typically contain 50–70% honey with added corn syrup, maltodextrins, and processing aids such as anticaking and drying aids. "Creamed" or "whipped" honey has been processed by controlled crystallization to give a smooth, spreadable consistency.

Starch-Derived Sweeteners

Starch is the major energy reserve for many plants, and contemporary agriculture has succeeded in efficiently making cornstarch available as an inexpensive food item. The creativity of chemists and engineers has exploited this material to produce a wide variety of food ingredients with different chemical, physical, and functional properties.

Corn Syrup

Water-insoluble starch granules can be solubilized with mechanical stirring and heat. Traditionally, the solubilized starch was treated with dilute hydrochloric acid under pressure to produce glucose, maltose, and other glucose oligosaccharides. Today, a milder treatment using amylase enzymes is used for hydrolysis. The glucose content of these syrups can vary from 20% to 90%. **Dextrose equivalency (DE)** is a measure of the percentage of reducing sugars in a starch product. Thus, glucose has a DE of 100 and glucose syrups will have a minimum DE of 20.

High-Fructose Corn Syrup (HFCS)

Conversion of glucose to fructose with the enzyme D-xylose isomerase was first developed in 1957, with technological improvements in the 1960s. In the 1970s, HFCS was widely adopted in the United States for use in soft drinks and other processed foods. Glucose syrup will be converted almost completely to fructose, which is then blended with glucose syrup to give syrups containing 42%, 55%, and 95% fructose. The most widely used HFCS has a fructose content of 55%. The sweetness properties of HFCS are substantially greater than the glucose syrups and very similar to other invert syrups.

Crystalline Fructose

Fructose has a higher solubility than other sugars. Highly concentrated solutions of fructose can be made, which are crystallized with difficulty. Industry has succeeded in producing crystalline fructose (99+% pure fructose), which is available for use as an alternative to HFCS and sucrose.

Maltodextrins

Maltodextrins are another product of corn syrup manufacture, being glucose oligosaccharides consisting of 3–20 glucose units. Another way of stating their size is that they have a **degree of polymerization (DP)** of 3-20. Conversely, they have a DE of 33 to 5. Thus, maltodextrins are a family of products having a wide range of functional properties and food applications. Their sweetness will vary inversely with their DP.

Inulin Syrup

Inulin is a polysaccharide found typically in the rhizomes and roots of plants, such as onions, Jerusalem artichokes, chicory, etc. It consists primarily of fructose, with inulin from some sources containing some glucose units. Hydrolyzed inulin syrup from Jerusalem artichoke has a fructose:glucose ratio of 2.80 (Beckers et al. 2008).

Sugar Composition of Fruits and Fruit Juices

Different fruits vary considerably with respect to sugar content. Grapes are relatively high in total sugar content, making the fruit very susceptible to fermentation, which was discovered centuries ago. Cranberries and lemons are both low in sugar and high in acid. Sugar needs to be added to their juices to render them palatable. Refractometry is a convenient means of measuring sugar content of fruits and fruit juices. The refractive index of aqueous sugar solutions varies linearly with concentration, and hand-held refractometers that can be used in the field and processing plant are calibrated in terms of percent soluble solids, or °Brix. °**Brix** is essentially the percent sugar by weight. A solution of 10 g of sugar in 90 g of water will be a 10.0 °Brix solution. The °Brix of wine grapes will be carefully monitored in the vineyard to help determine time of harvest. Wine will typically be made from grapes ranging from 21 to 25 °Brix. Sugar content of a given fruit commodity can vary considerably with variety, maturity, growing region, and farming practices, such as irrigation. For example, a

Table 2.2   Sugar Composition[a] of Selected Fruit Juices

Fruit	°Brix[b]	%Fru[c]	%Glu[c]	Fru:Glu	%Suc[c]	%Sorb[c]
Apple[d]	11.5	52	19	2.7	24	5
Blackberry[e]	10.0	48	48	1.0	4	
Blueberry[f]	10.0	49	49	1.0	2	
Boysenberry[e]	10.0	47	48	0.99	5	
Cherry, dark sweet[g]	20.0	43	47	0.92	2	8
Cherry, red sour[h]	14.0	37	35	1.06		27
Chokeberry (*Aronia melanocarpa*)[i]		25	27	0.95	tr	48
Cranberry[j]	7.5	23	78	0.30		
Grape[g]	16.0	47	48	0.98	4	
Grapefruit[k]	10.0	34	31	1.10	35	
Kiwi[l]	15.4	44	48	0.92	8	
Lemon[k]	4.5	33	43	0.77	24	
Orange[k]	11.8	28	26	1.08	46	
Peach[g]	10.5	12	11	1.09	67	10
Pear[g]	12.0	54	14	3.86	14	18
Pineapple[m]	12.8	25	22	1.14	53	
Plum[g]	14.3	20	33	0.60	33	14
Pomegranate[n]	16.0	46	54	0.86		
Prune[m]	18.5	25	46	0.54		30
Red Raspberry[o]	9.2	50	46	1.09	4	
Strawberry[g]	8.0	41	43	0.95	17	

[a]Expressed as percent of fructose + glucose + sucrose + sorbitol.
[b]FDA °Brix standards for single-strength fruit juices (Code of Federal Regulations 2010).
[c]Abbreviations: Fru, fructose; Glu, glucose; Suc, sucrose; Sorb, sorbitol; tr, trace.
Sources: [d]Mattick and Moyer 1983; [e]Fan-Chiang and Wrolstad 2010; [f]Ayaz et al. 2001; [g]Wrolstad and Shallenberger 1981; [h]Gao et al. 2003; [i]Durst et al. 1996; [j]Coppola and Starr 1988; [k]Dillon 1995; [l]Lodge and Perera 1992; [m]Pilando and Wrolstad 1992; [n]Sugiyama et al. 1992; [o]Durst et al. 1995.

large 3-year study determining the composition of different varieties of apples from different growing regions in the United States was conducted at Cornell University from 1979 to 1981 (Mattick and Moyer 1983). The range in °Brix was 9.8–16.9° with a mean of 12.74° for this large sample ($n = 93$) study. The FDA has established single-strength °Brix values for many fruit juices for purposes of commercial trade and label declaration of the amounts of fruit juice in juice drinks and blended fruit juices (Code of Federal Regulations 2010). The single-strength °Brix standards for a selection of fruit juices are listed in Table 2.2.

Anecdote

In 1991, Joe and Jean Briggs, owners of Briggs Orchard, an organic apple orchard located in Arizona's high altitude Bonita Valley, signed a contract with a Japanese firm, Made In Nature, Inc., for purchase of their entire next season's crop. In 1992, the weather was unseasonably cold, and the apples were slow to ripen. Joe Briggs repeatedly contacted Made In Nature, advising the company that the apples were ready for harvest. Made In Nature insisted that the fruit was immature, as it was of low °Brix. Briggs stated that depletion of starch by the starch iodine test was a more reliable indication of maturity, and the fruit was past the optimum harvest date. Eventually, the apples were allowed to rot. Litigation took place, and the defense of Made In Nature for nonpayment of the fruit was that the apples were below the single-strength standard of 11.5 °Brix, and the derived juice could not legally be considered to be apple juice. (Not coincidentally, the price of apple juice concentrate dropped from a high of $13.00–$14.00/gallon in 1991 to $8.00–$8.50/gallon in 1992). The judge's decision was that Made In Nature did owe a substantial payment to Briggs Orchard.

Table 2.2 also lists the glucose, fructose, sucrose, and sorbitol content of fruit juices expressed as a percent of total sugars (glucose + fructose + sucrose + sorbitol). An estimate of the actual quantity of individual sugar in single-strength juice can be obtained by multiplying the percent sugar by the standard °Brix. The different fruits exhibit distinctive patterns that are characteristic of the fruits. Glucose and fructose are the major sugars in all these fruits, and many fruits (blackberries, blueberries, cherries, oranges, raspberries) have an invert pattern of nearly equal amounts of glucose and fructose. This is readily seen by the glucose:fructose ratio, which is also listed in Table 2.2. Apples and pears accumulate more fructose than glucose, and cranberry is unusual in that it accumulates more glucose than fructose. Sucrose tends to be present in small quantities or is nonexistent. Sucrose values for fruit juices tend to be highly variable because of acid or enzymic hydrolysis that can occur during processing and storage. Durst et al. (1995) showed that, although sucrose was present in small and varying amounts in red raspberry fruits, it was detected in only trace amounts, if at all, in raspberry juice concentrates. Some fruits, such as apples, pears, cherries, peaches, and plums, accumulate significant amounts of sorbitol. Sorbitol can serve as a marker compound for these juices, when they are undeclared and illegally used to extend more expensive nonsorbitol-containing fruit juices (Wrolstad et al. 1981). Trace amounts of sorbitol in juices from nonsorbitol-containing fruits, such as strawberries and raspberries, may indicate contamination from other fruit juices. Another possibility, however, could be that the source was commercial juice-processing enzymes, which sometimes are in solutions that contain sorbitol as a protective osmotic agent (Durst et al. 1995). Sorbitol is nonfermentative; hence, fruit wines from sorbitol-containing fruits, such as pears, will have residual sweetness even when fermented to dryness. Chokeberries (*Aronia melanocarpa*) are unusual in that some genotypes contain as high as 55% sorbitol of total sugars (Durst et al. 1996). This may be useful to some diabetics, as sorbitol is more slowly absorbed. (see Chapter 9). The proportions and ratios of these sugars are less variable than the absolute content of individual sugars (Wrolstad and Shallenberger 1981). The glucose:fructose ratio and percent sorbitol content are particularly useful indices in fruit juice authenticity investigations (Wrolstad et al. 1981).

Anecdote

The Beech-Nut adulterated apple juice for babies scandal that surfaced in 1988 is a landmark case in food law (Wrolstad 1991; Wrolstad and Durst 2007). FDA took action against Beech-Nut, who had purchased apple juice concentrate at a 20% cost advantage from Food Complex, Inc. and Universal Juice Co. that was purported to be 100% apple juice. In fact, the concentrate was formulated from invert beet sugar syrup, corn syrup, synthetic malic acid, caramel coloring, and imitation apple juice flavoring. Detection of synthetic (DL) malic acid by a combination of enzymic and HPLC analyses was critical to the case. Additional chemical evidence for the juice not being authentic was an invert sugar pattern and the absence of sorbitol. The sugar profile for authentic apple juice was firmly established in the literature at that time, and Beech-Nut should have been aware that the concentrate was not authentic. More than $2 million in fines was paid by Beech-Nut, and company executives were fined and given prison sentences. Losses in sales to Beech-Nut were estimated to be $25 million. Gerber Products Co., a competitor of Beech-Nut, unjustly suffered losses in sales from consumer reaction to the scandal.

Vocabulary

°Brix—Percent sugar by weight in an aqueous solution
Degree of polymerization (DP)—Number of sugar units in a molecule
Dextrose equivalency (DE)—A measure of the percentage of reducing sugars in a starch product; the percentage of reducing sugars in a starch hydrolysate on a dry weight basis (e.g., glucose = 100 and starch = 0)

References

Ayaz FA, Kadioglu A, Bertoft E, Acar C, Turna I. 2001. Effect of fruit maturation on sugar and organic acid composition in two blueberries (*Vaccinium arctostaphylos* and *V. myrtillus*) native to Turkey. *N Z J Crop Hort Sci* 29:137–141.

Coppola ED, Starr MS. 1988. Determination of authenticity and percent juice of cranberry products. In: Nagy S, Attaway JA, Rhodes ME, editors. *Adulteration of fruit juice beverages*. New York: Marcel Dekker, pp. 139–174.

Beckers M, Grube M, Upite D, Kaminska E, Danilevich A, Viesturs U. 2008. Inulin syrup from dried Jerusalem artichoke. *LLU Raksti* 21:116–121.

Code of Federal Regulations. 2010. 21 CFR 101.30: Percentage juice declaration for foods purporting to be beverages that contain fruit or vegetable juice. Available from: http://cfr.vlex.com/vid/101-percentage-juice-purporting -contain-19705632. Accessed October 18, 2010.

Dillon A. 1995. Appendix II: Fruit juice profiles. In: Nagy S, Wade RL, editors. *Methods to detect adulteration of fruit juice beverages, volume I*. Auburndale, FL: AgScience, pp. 359–438.

Durst RW, Wrolstad RE, Krueger DA. 1995. Sugar, nonvolatile acid, $^{13}C/^{12}C$ ratio, and mineral analysis for determination of the authenticity and quality of red raspberry juice composition. *J Assoc Off Anal Chem* 78:1195–1204.

**Durst RW, Wrolstad RE, Mills JA. 1996. Unpublished data.

Fan-Chiang H-J, Wrolstad RE. 2010. Sugar and nonvolatile acid composition of blackberries. *J Assoc Off Anal Chem* 93:956–965.

Gao Z, Maurousset L, Lemoine R, Yoo S-D, van Nocker S, Loescher W. 2003. Cloning, expression, and characterization of sorbitol transporters from developing sour cherry fruit and leaf sink tissues. *Plant Physiol* 131:1566–1575.

Lodge N, Perera CO. 1992. Processing of Kiwifruit. *NZ Kiwifruit* July 1992:14–15.

Mattick LR, Moyer JC. 1983. Composition of apple juice. *J Assoc Off Anal Chem* 66:1251–1255.

Pancoast HM, Junk WR. 1980. *Handbook of sugars, 2nd ed*. Westport, CT: AVI Publishing.

Pilando LS, Wrolstad RE. 1992. Compositional profiles of fruit juice concentrates and sweeteners. *Food Chem* 44:19–27.

Sugiyama N, Roemer K, Bunemann G. 1992. Sugar patterns of exotic fruits on a German fruit market. *Acta Horticult* 321:850–855.

[USDA] United States Department of Agriculture. 1985. *United States standards for grades of extracted honey*. Washington, D.C.: USDA.

USDA. 2009. *USDA national nutrient database for standard reference, release 22.* Available from: http://www.ars.usda.gov/nutrientdata. Accessed September 10, 2010.

Wrolstad RE. 1991. Ethical issues concerning food adulteration. *Food Technol* 45:108, 112, 114, 116–117.

Wrolstad RE, Cornwell CJ, Culbertson JD, Reyes FGR. 1981. Establishing criteria for determining the authenticity of fruit juice concentrates. In: Teranishi R, Barrera-Benitez H, editors. *Quality of selected fruits and vegetables of North America*. ACS Symposium Series 170. Washington, D.C.: American Chemical Society, pp. 77–93.

Wrolstad RE, Durst RW. 2007. Fruit juice authentication: what have we learned? In: Ebeler SE, Takeoka GR, Winterhalter P, editors. *Authentication of food and wine.* ACS Symposium Series 952.Washington, D.C.: American Chemical Society, pp. 147–163.

Wrolstad RE, Shallenberger RS. 1981. Free sugars and sorbitol in fruits: a compilation from the literature. *J Assoc Off Anal Chem* 64:91–103.

3 Reactions of Sugars

Introduction

The reactions of sugars that are critical in food science applications can be organized into a few general types of reactions that are heavily influenced by pH. Although there are a few exceptions, most of these reactions occur under the relatively mild conditions that prevail during processing and storage. This chapter will not address reactions of chemical synthesis and sugar structure proof. The monograph *Carbohydrates—The Sweet Molecules of Life* by R. V. Stick (2001) is recommended for its excellent treatment of sugar synthesis and basic sugar reactions.

Mutarotation

Mutarotation can be defined as the change in optical rotation that is observed when a reducing sugar is dissolved in water, due to the formation of different tautomeric forms. A sugar crystal will consist of molecules having a specific anomeric ring form (furanose or pyranose with α- or β-configuration). Upon dissolution, ring opening (hydrolysis) and

Food Carbohydrate Chemistry, First Edition. Ronald E. Wrolstad.
© 2012 John Wiley & Sons, Inc. Published 2012 by John Wiley & Sons, Inc.

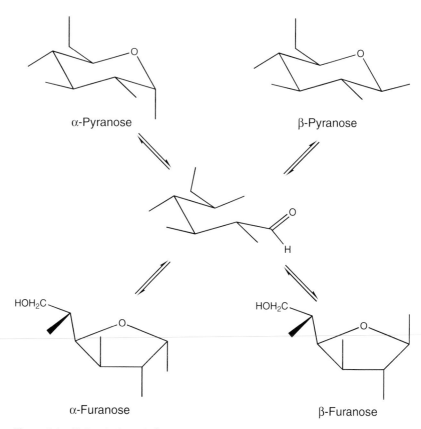

α-Pyranose   β-Pyranose

α-Furanose   β-Furanose

Figure 3.1   Mutarotation of glucose.

subsequent ring closure will occur, producing the α- and β-pyranose and
α- and β-furanose forms. These forms have different chemical and physical properties (e.g., optical rotation, solubility, chemical reactivity, relative
sweetness, etc.). Figure 3.1 shows the five forms of D-glucose that will the-
oretically exist in solution. For glucose, only the α- and β-pyranose forms
exist in significant amounts. α-D-Glucopyranose has an "initial" optical
rotation of +112°, whereas β-D-glucopyranose has an "initial" rotation of
+19°. The amounts of the tautomers will be governed by conformational
free energy. β-D-Glucopyranose has the greatest stability and predominates by being present at 63.6% at equilibrium at 20°C. Table 3.1 gives
the distribution of selected D-sugar tautomers at equilibrium in water
solution.

   Glucose is classified as undergoing "simple" mutarotation since,
for practical purposes, only two tautomers are present. This permits

Table 3.1   Distribution of D-Sugar Tautomers at Equilibrium in Aqueous Solution[a]

Sugar	Mutarotation Type	Temp (°C)	α-Pyranose (%)	β-Pyranose (%)	α-Furanose (%)	β-Furanose (%)	Open-chain (%)
Glucose	Simple	20	36.4	63.6	—	—	—
	Simple	31	38	62	0.5	0.5	0.002
Galactose	Complex	20	32	63.9	1	3.1	
	Complex	31	30	64	2.5	3.5	0.02
Fructose	Complex	27	—	75	4	21	
	Complex	31	2.5	65	6.5	25	0.8
Xylose	Simple	20	34.8	65.2	—	—	
	Complex	31	36.5	58.5	6.4	13.5	0.05

[a]Modified from Cui 2005.

calculation of the amounts of α- and β-pyranose algebraically.

Let X = % α-D-glucopyranose at time t;

then 1 − X = % β-D-glucopyranose.

$[\alpha]_D$ α-D-glucopyranose = +112°; $[\alpha]_D$ α-D-glucopyranose = +19°;

$[\alpha]_D$ at equilibrium = +52° X (112) + (1 − X)(19) = 52

112 X + 19 − 19X = 52; 93X = 33;

X = 33/93 = 33%α-D-gluopyranose; 66% β-D-glucopyranose   (3.1)

Note that in Table 1.3 the distribution of sugar tautomers is temperature-dependent. As the temperature is increased from 20°C to 31°C, the amounts of higher energy α-pyranose, furanose, and open-chain forms of xylose increase so that "complex" mutarotation is the appropriate classification. (Complex mutarotation denotes that more than 3 tautomers are present in significant amounts, making calculation of quantities from optical rotation measurements difficult, if not impossible.) This temperature effect can have some very important ramifications for foods that are consumed hot, ice-cold, or at room temperature. Figure 3.2 shows the effect of temperature on the relative sweetness of four different sugars. Although fructose is much sweeter than glucose or sucrose (sucrose = 100, the reference for sweetness) at low temperatures, it is less sweet than sucrose at 60°C. This is because higher amounts of β-D-fructopyranose, which is intensely sweet, exist at the colder temperature. Larger amounts of the open-chain form exist at higher temperatures. Many sugar reactions, such as reaction of reducing sugars with amino compounds in the Maillard reaction (Chapter 4), will occur at a more rapid rate proportional to the amount of sugar in the open-chain from. In general, pentose sugars have higher

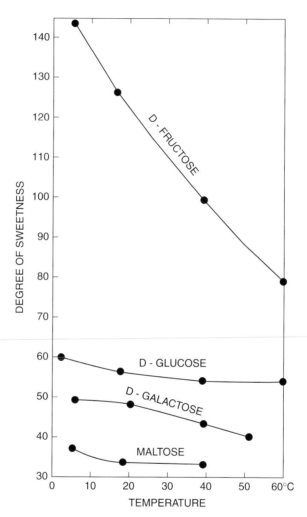

Figure 3.2   Effect of temperature on the relative sweetness of sugars.
Source: **Shallenberger and Birch 1975.**

amounts in the open-chain form than hexose sugars and are more reactive in Maillard browning.

Ring opening can be catalyzed by both acid and base (Figure 3.3). Under acidic conditions, a proton will attack the ring oxygen with subsequent ring opening. In alkaline conditions, the hydroxide ion will initiate ring opening through removal of the proton of the anomeric hydroxyl group. The mechanism was substantiated by conducting experiments in H_2O^{18}. There was no incorporation of O^{18} into the sugar molecule; therefore,

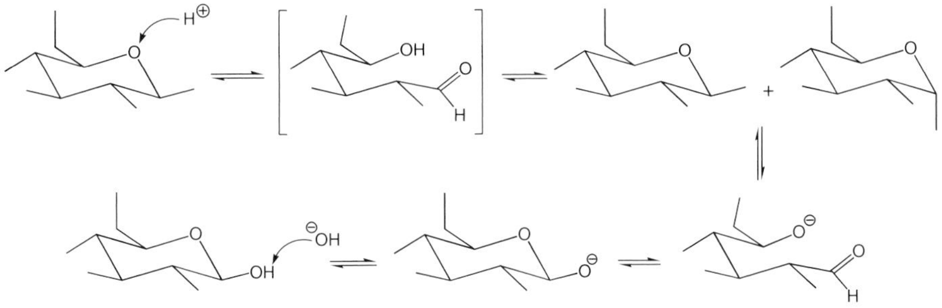

Figure 3.3 Mechanism for acid and base catalysis of mutarotation.

hydroxide ion did not displace the anomeric hydroxyl group (Capon 1969). Mutarotation will occur under neutral conditions because water can function as either an acid or a base. The rate of reaction will occur much faster, however, in the presence of H+ or OH−. The Appendix includes a laboratory experiment that demonstrates mutarotation and the optical properties of sugar solutions using a polarimeter.

Oxidation of Sugars

Aldose sugars can be readily oxidized to aldonic acids (e.g., D-glucose to D-gluconic acid). There are a number of qualitative and quantitative tests used to detect and measure reducing sugars that are based on sugar oxidation. In the Fehling's test, an alkaline solution of copper$^{(++)}$sulfate oxidizes the aldose to an aldonic acid, the cupric ion being reduced to cuprous hydroxide that precipitates as red-colored cuprous oxide.

$$2 \text{ Cu (OH)}_2 + \text{Aldose sugar} \rightarrow \text{Aldonic Acid} + \text{Cu}_2\text{O} + \text{H}_2\text{O}$$

Reaction 3.2

Under these conditions, fructose isomerizes to aldose sugars and gives a positive test. (Not all ketoses will give a positive test, however.) A laboratory experiment is described in the Appendix that uses the Fehling's test for determination of whether a selection of carbohydrate materials is reducing or nonreducing. The described protocol is qualitative. It can be modified to be quantitative and is frequently used in the wine industry to determine the amount of reducing sugars (Zoecklein et al. 1995; refer to Chapter 6). There are other similar methods for determining reducing sugars that are based on sugar oxidation and metal ion reduction, for example, Somogyi-Nelson, Munson-Walker, etc. (BeMiller 2010).

Glucose oxidase specifically oxidizes D-glucopyranose to D-gluconic acid. Enzyme test kits are available that permit measurement of D-glucose content in the presence of other sugars in foods and biological systems (see Chapter 6). Another food application is the use of a combination of glucose oxidase and catalase in glucose-containing foods for removal of either glucose or oxygen (Parkin 2008). It is used to deplete egg white of glucose before spray drying to minimize the Maillard reaction (see Chapter 4). It is also effective for removal of oxygen in citrus fruit juices and within sealed package systems.

$$\text{Glucose} + O_2 \xrightarrow{\textit{Glucose oxidase}} \text{D - Gluconic acid} + H_2O_2 \xrightarrow{\textit{catalase}} H_2O + \tfrac{1}{2} O_2$$

Reaction 3.3

Glycoside Formation

Reducing sugars will react with alcohol under anhydrous and acidic conditions to form acetals or ketals, which are nonreducing. This is illustrated by the reaction of glucose and methanol to give methyl-α-D-glucopyranoside and methyl-β-D-glucopyranoside.

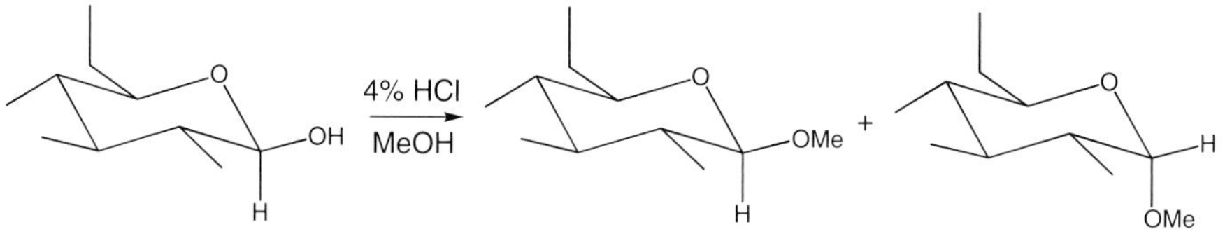

Reaction 3.4

The traditional process for corn syrup manufacture (see Chapter 2) gives reaction conditions suitable for glycoside formation during the latter stages of processing (high concentration of glucose, heat, and acid). **Reversion sugars** are oligosaccharides that are formed from acid-catalyzed condensation of glucose.

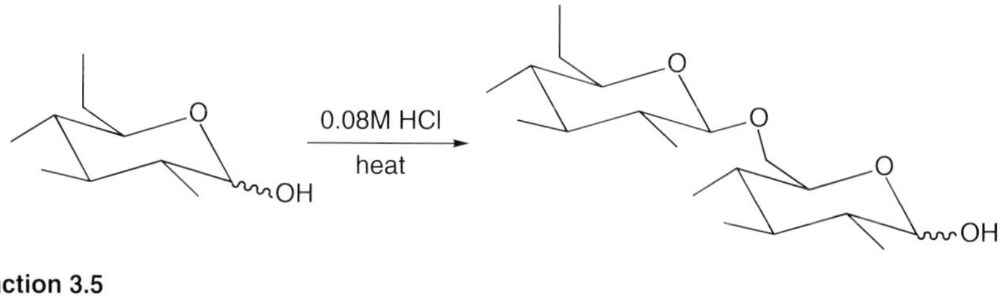

Reaction 3.5

Gentiobiose (6-O-β-D-glucopyranosyl-β-D-glucopyranose), which has a bitter taste, is formed in significant quantities. The condensation reaction is nonspecific, and although β-1-6 predominates, 1-4, 1-3, and 1-2 oligosaccharides are also formed in decreasing amounts, respectively (Shallenberger and Birch 1975). There will be a mixture of α and β glycosidic linkages. Condensation between the anomeric hydroxyls to produce trehalose (α-D-glucopyranosyl-α-D-glucopyranoside) does not occur because the mechanism requires both a hydroxyl that is a strong proton donor (anomeric OH) and one that is also a strong proton acceptor. The hydroxyl functions on carbons 2, 3, 4, and 6 will serve as a proton acceptor, but the anomeric hydroxyl is ineffective in this regard.

Glycosans (anhydro sugars), which are intramolecular glycosides and therefore nonreducing, can be formed from aldose sugars subjected to acid and heat. When glucose and idose are subjected to identical reaction conditions, levoglucosan will be produced in only 0.3% yield, whereas idose will produce the corresponding glycosan in 80% yield. In order for the reaction to occur, the sugar must be in the 1-C conformation, which is the preferred conformation for idose and almost nonexistent for glucose. The important role of conformation for this reaction is clearly demonstrated with molecular models.

Reaction 3.6

Polydextrose, also marketed as Litesse^{TM}, Sta-lite^{TM}, and Trimcal^{TM} is a water-soluble, reduced-calorie bulking agent approved for food use. It is classified as soluble fiber and is used to increase the nondietary fiber content of foods through partial replacement of sugar, starch, and fat in a wide range of food products. The polymer is manufactured by pyrolysis of glucose, sorbitol, and citric acid with the proportions of 89 parts glucose, 10 parts sorbitol, and 1 part citric acid. The product is randomly cross-linked, with 1-6 linkages predominating and a mixture of α and β linkages. The carbohydrase enzymes in the intestinal mucosa will not hydrolyze many of these linkages. Hence, much of the material is not digested, absorbed, or used. The caloric value is approximately 1 cal/g.

The formation of trimethylsilyl derivatives of sugars permits analysis of anomeric and ring forms of sugars by gas-liquid chromatography. The analysis is extremely sensitive. In the derivation reaction shown in Reaction 3.7, a trimethylsily glycoside is formed at the anomeric carbon, with trimethylsily ethers being formed at the other hydroxyls (see Chapter 6).

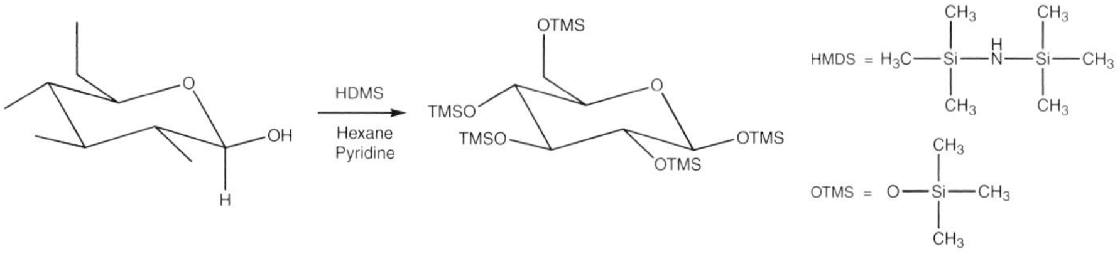

Reaction 3.7

Acid Catalyzed Sugar Reactions

Attack by a proton on the ring oxygen of a reducing sugar results in ring opening and mutarotation as previously described (Figure 3.3). Hydrolysis of the glycosidic linkage in reducing and nonreducing sugars is catalyzed by protonation of the exocyclic OR group. A carbonium ion is formed that abstracts a hydroxyl group from water to form free sugars (Wong 1989).

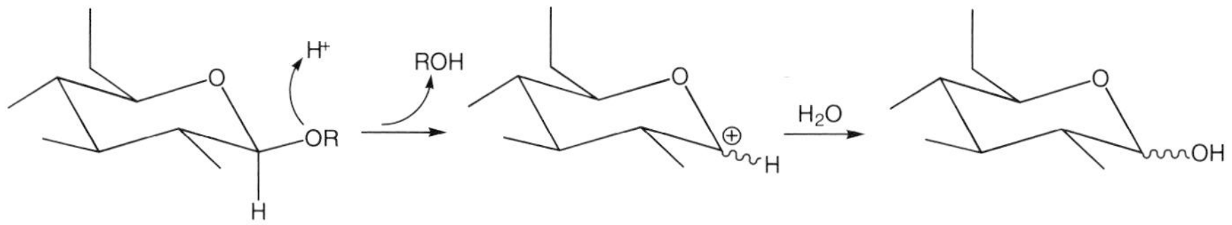

Reaction 3.8

Hydrolysis rates are predictable with furanose linkages being hydrolyzed much faster than pyranose linkages. α-Glycosidic linkages are hydrolyzed faster than β. Hydrolysis rates of glycosidic linkages occur in the following decreasing order: 1-2 > 1-3 > 1-4 > 1-6. Therefore, one can predict that sucrose (α-furanose) will be hydrolyzed much easier than maltose (α-1-4), which undergoes hydrolysis at a faster rate than lactose or cellobiose (both β-1-4). It also explains why gentiobiose (β-1-6) accumulates in corn syrup manufacture.

Aldose sugars will undergo acid-catalyzed dehydration to form deoxy sugars and furfurals. Figure 3.4 shows the mechanism for dehydration of

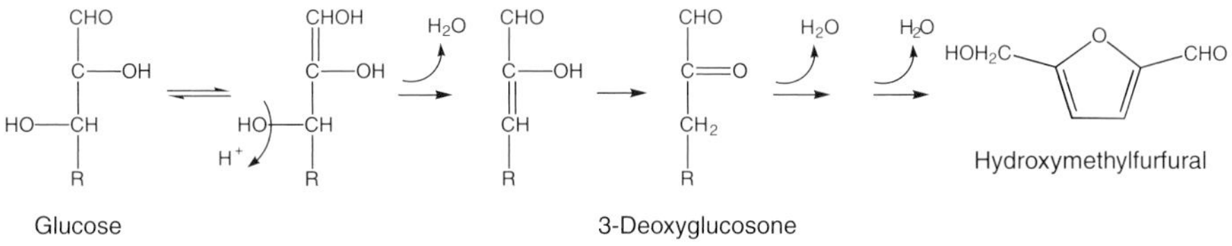

Figure 3.4   **Dehydration of glucose via 2-3-enolization to form 3-deoxyglucosone.**

glucose through 2-3-enolization to form 3-deoxyglucosone. Initial proton attack at C-3 results in hydroxyl removal at C-3, and subsequent proton removal at C-2 gives the 2-3-enol form; ketol-enol rearrangement produces 3-dexoyglucosone. Further dehydration occurs down the chain with subsequent ring closure to produce hydroxymethylfurfural from aldose sugars and furfural from pentoses.

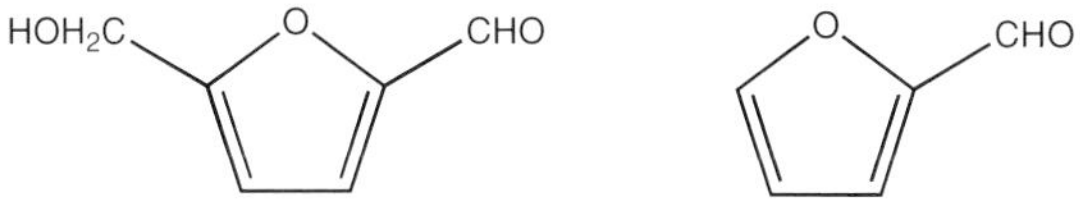

Structure 3.9   **Hydroxylmethylfurfural and Furfural**

Furfurals have flavor significance, and they are reactive in polymerization reactions. Another laboratory application is for visualization of sugars in different chromatographic systems. For example, a thin-layer chromatogram of separated sugars can be sprayed with an acidic solution and heated to produce furfurals that are subsequently reacted with amines to give a colored complex (Lewis and Smith 1969).

Alkaline-Catalyzed Sugar Reactions

Mutarotation can be alkaline catalyzed whereby hydroxide ion initiates proton removal from the anomeric hydroxyl, followed by ring opening (Figure 3.3). In the case of nonreducing sugars, there is no anomeric hydroxyl group. Therefore, under alkaline conditions, hydrolysis will not take place.

A practical application is the use of lime in sugar refining to render the extract alkaline so that hydrolysis of the acid-labile α-furanose linkage (sucrose inversion) does not take place.

Reaction 3.10   (Base-catalyzed mutarotation of glucose vs. no reaction for acetal.)

Reducing sugars will undergo 1-2-enolization to form different isomers (Figure 3.5). The reaction is favored in alkali but can also occur under acidic conditions. This is one reason that Fehling's reagent is alkaline (refer to the Fehling's test for reducing sugars in the Appendix). The enediol is readily oxidized by metal ions, and, in addition, acid hydrolysis of labile nonreducing sugars, such as sucrose, is prevented. One application of this reaction is to convert the glucose moiety of lactose (4-O-β-D-galactopyranosyl-D-glucose) to lactulose (4-O-β-D-galactopyranosyl-D-fructose). The reaction is catalyzed by sodium aluminate.

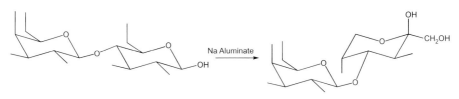

Reaction 3.11

Lactulose has higher water solubility than lactose, and it is somewhat sweeter than lactose, 0.5× sucrose versus 0.4× sucrose. Lactulose is not hydrolyzed by lactase, not absorbed, and noncaloric. It is used as a laxative but is not approved for food use.

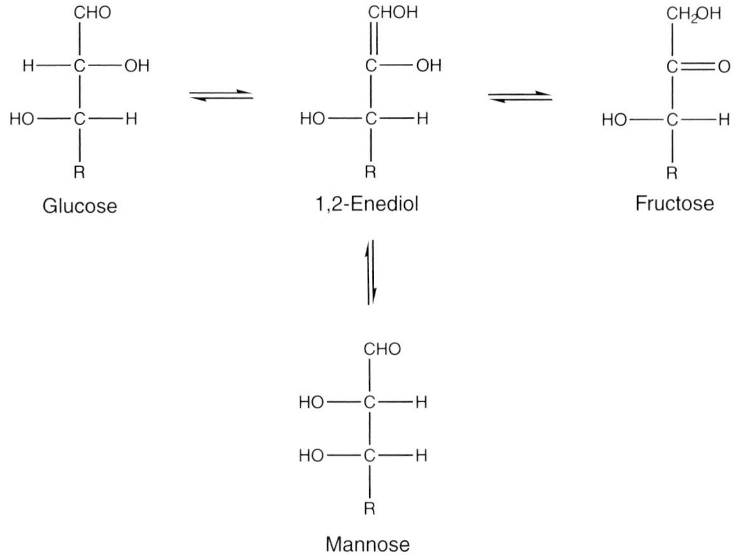

Figure 3.5   Isomerization of glucose to mannose and fructose via 1-2-enolization.

Summary

Key reactions in carbohydrate chemistry are influenced by conditions of acidity and alkalinity. Table 3.2 summarizes those effects.

Similarly, many carbohydrate reactions can be categorized as being reactions of reducing and nonreducing sugars. The properties of reducing and nonreducing sugars can be summarized as shown in Table 3.3.

Figure 3.6 summarizes several key reactions of sugars. Transformation of β-D-glucopyranose (compound 2) to α-D-glucopyranose (compound 1) via the open-chain form (compound 5) is mutarotation. Transformation of compound 2 to compound 3 is conformational change (C-1 to 1-C). Conversion of compound 3 to compound 4 is formation of the

Table 3.2   Key Carbohydrate Reactions as Influenced by Acid and Alkali

Reaction	Favored in Acid	Favored in Alkali
Mutarotation	+	+
Hyrolysis of hemiacetals and hemiketals	+	+
Hydrolysis of acetals and ketals	+	-
Polymerization	+	
Dehydration	+	
Enolization	+ (2–3)	+ (1–2)

Table 3.3   Properties of Reducing and Nonreducing Sugars

Reducing Sugars	Nonreducing Sugars
Will reduce metal ions	Not oxidized by metal ions
Undergo mutarotation	Will not undergo mutarotation
Undergo acid hydrolysis	Undergo acid hydrolysis
Hydrolyzed by alkali	Not hydrolyzed by alkali
Hemiacetals and hemiketals	Acetals and ketals
-ose suffix	-ide suffix
Condense with amines in Maillard reaction	Nonreactive in Maillard reaction

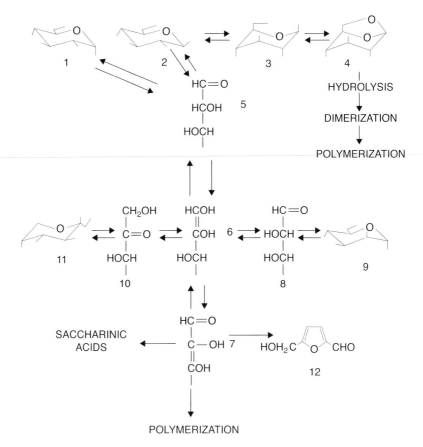

Figure 3.6   Reactions of reducing sugars in solution.
Source: **Shallenberger and Birch 1975.**

anhydrosugar, levoglycosan. Conversion of compound 5 to 6 is 1-2-enolization, with subsequent isomerization to fructose (compounds 10 and 11) and mannose (compounds 8 and 9). Acid catalyzed 2-3-enolization (compound 7) leads to formation of hydroxymethylfurfural (compound 12) and formation of polymers and saccharinic acids.

Vocabulary

Mutarotation—The change in optical rotation that is observed when a reducing sugar is dissolved in water due to the formation of different tautomeric forms. "Simple" mutarotation describes the case when only two tautomers exist, and "complex" mutarotation describes the existence of three or more tautomers.

Reversion sugars—Oligosaccharides formed from acid-catalyzed condensation of glucose

Glycosans—Anhydro sugars, which are intramolecular glycosides and therefore nonreducing

References

BeMiller JN. 2010. Carbohydrate analysis. In: Nielsen SS, editor. *Food analysis, 4th ed.*, New York: Springer, pp. 147–177.

Capon B. 1969. Mechanism in carbohydrate chemistry. *Chem Rev* 69:407–498.

Cui SW. 2005. *Food carbohydrates: chemistry, physical properties, and applications.* Boca Raton, FL: CRC Press, Taylor & Francis Group.

Lewis BA, Smith F. 1969. Sugars and derivatives. In: Stahl E, editor. *Thin-layer chromatography: a laboratory handbook, 2nd ed.*, New York: Springer-Verlag, pp. 807–837.

Parkin KL. 2008. Enzymes. In: Damodaran S, Parkin KL, Fennema OR, editors. *Fennema's food chemistry, 4th ed.*, Boca Raton, FL: CRC Press, pp 331–435.

Shallenberger RS, Birch GG. 1975. Sugar chemistry. Westport, CT: AVI Publishing.

Stick RV. 2001. *Carbohydrates: the sweet molecules of life.* New York: Academic Press.

Wong DWS. 1989. *Mechanism and theory in food chemistry.* New York: Van Nostrand Reinhold.

Zoecklein BW, Fugelsang KC, Gump BH, Nury FS. 1995. *Wine analysis and production.* New York: Chapman and Hall.

4 Browning Reactions

Food Carbohydrate Chemistry, First Edition. Ronald E. Wrolstad.
© 2012 John Wiley & Sons, Inc. Published 2012 by John Wiley & Sons, Inc.

Introduction

Maillard browning and lipid oxidation are arguably the two most important topics in food chemistry. Neither is a single chemical reaction; rather, they consist of a series of complex reactions that follow different routes under changing conditions. A French chemist, Louis Maillard, was the first to describe the formation of brown pigments from heating a solution containing glucose and lysine (Maillard 1912). Inspection of the annual indexes for the *Journal of Agricultural and Food Chemistry* and the *Journal of Food Science* will predictably have numerous articles on nonenzymatic browning from the time the journals were first published. Science Citation Index lists 610 articles on the topic of Maillard browning in the time period from 1970 to the present (SCI 2010). The topic has had sufficient interest to establish the International Maillard Reaction Society (IMARS), which has been organizing symposia on the Maillard reaction every 3–4 years since 1982 (IMARS 2010).

Nonenzymatic browning reactions are responsible for the desirable colors of many baked goods and beverages. In beer production, for example, it is critical that the color intensity be controlled so that consumer expectations for the appearance of lager, ale, and stout are met (Shellhammer and Bamforth 2008). The formation of brown pigments during processing and storage are often a quality defect that needs to be minimized. Flavors are also generated through nonenzymatic browning that may be desirable, for example, the flavors of roasted coffee and the South American dessert dulce de leche. The reaction has nutritional significance in that it can reduce protein quality from the degradation of lysine and other essential amino acids. Formation of mutagens and harmful compounds, such as acrylamide, are reasons for continued active research on the topic. Food technologists will monitor reaction products during processing and storage and establish purchase and quality specifications on the basis of the extent of their formation.

This chapter will not be a comprehensive review of the topic. For such information, the following references are suggested: Waller and Feather 1983; Ames 1998; Martins et al. 2000; Nursten 2005; Gerrard 2006. Recent reviews and symposia reflect the shift in research interest to antioxidant properties (Manzocco et al. 2001) and health risks and benefits (O'Brien et al. 1998; Baynes et al. 2005; Somoza 2005). An effort will be made to present the key reactions and mechanisms involved in nonenzymatic browning, in order to have an understanding that will permit one to minimize undesirable effects and optimize and control desirable characteristics of the Maillard reaction.

Key Reactions in Maillard Browning

Introductory Comments

John Hodge, USDA chemist at the Wyndmor Regional Laboratory, did much to increase our understanding of Maillard browning through his research and reviews of the subject (Hodge 1953, 1967). Figure 4.1 is a modification of his figure (Hodge 1967) that has been reproduced in numerous books and review articles. It presents generalized structures of the types of compounds that appear as reducing sugars, and amino acids react sequentially to produce melanoidin pigments and flavor volatiles. Two routes are presented, one representing acidic reaction conditions and the other, a more alkaline environment.

Sugar-Amino Condensation

The initial reaction in Maillard browning is the condensation of a sugar and an amino compound to form a glycosyl amine, with loss of water (Reaction 4.1).

Reaction 4.1   Condensation of a reducing sugar with an amine to form a glycosyl amine.

The reaction involves nucleophilic attack of the amino group on the anomeric carbon. Condensation will be favored under concentrated conditions. A high pH with a greater proportion of the amino function in the free amine form will also favor the reaction. The mechanism is believed to involve formation of a Schiff's base, which is the product of condensation of aldehydes and amino compounds with elimination of water (Reaction 4.2).

Reaction 4.2   Condensation of an amino acid with an aldehyde to form a Schiff's base with elimination of water.

52

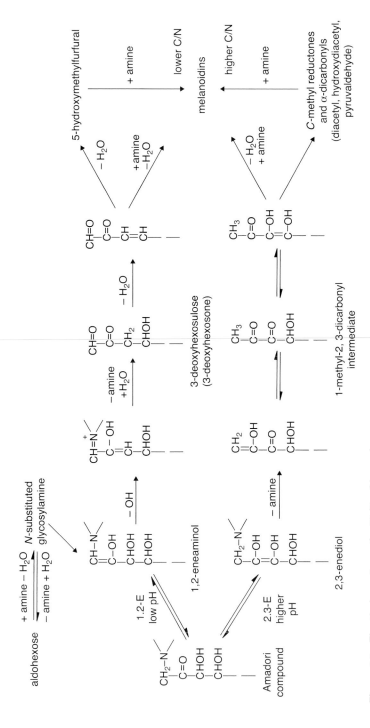

Figure 4.1 The Hodge scheme for Maillard browning.
Source: Eskin 1990, with permission.

Table 4.1   Comparison of Acylic Concentrations, Mutarotation Rates, and Browning Susceptibility of Selected Reducing Sugars[a]

Sugar	% Acyclic Form in H_2O	Relative to Glucose	Relative Mutarotation Rate	Browning Index[b]
Glucose	0.025	1	1	0.30
Mannose	0.064	2.7	1.7	0.42
Galactose	0.082	3.4	1.6	0.45
Xylose	0.17	7.1	3.2	1.30
Arabinose	0.28	11.7	4.8	0.96
Ribose	8.5	354	>500	2.04

[a] Adapted from Labuza and Schmidl 1986 and Hashiba 1982.
[b] E_{550nm} for 0.1M sugar and 0.1M glycine solutions (pH 5.0) heated at 120°C for 5 min.

The acyclic Schiff base intermediate is shown in Reaction 4.3.

Reaction 4.3   Formation of a Schiff's base intermediate with elimination of water.

Those sugars that exist in solution with a comparatively higher proportion in the acyclic form (refer to Table 3.1) will favor the reaction (Burton and McWeeny 1963). Table 4.1 compares the amounts of the acyclic form for selected sugars with their degree of browning.

The Amadori and Heyn's Rearrangements

An aldosyl amine will undergo rearrangement to form an amino-ketose. Figure 4.2 shows a mechanism for formation of 1-deoxy-1-glycino-β-D-fructose (referred to as the Amadori compound in Figure 4.1) from β-D-glucosyl-glycine.

The suggested mechanism involves ring opening, loss of water with formation of a Schiff base, formation of an enol through protonation of the amine, proton loss at carbon-2, enol-keto transformation, and ring closure.

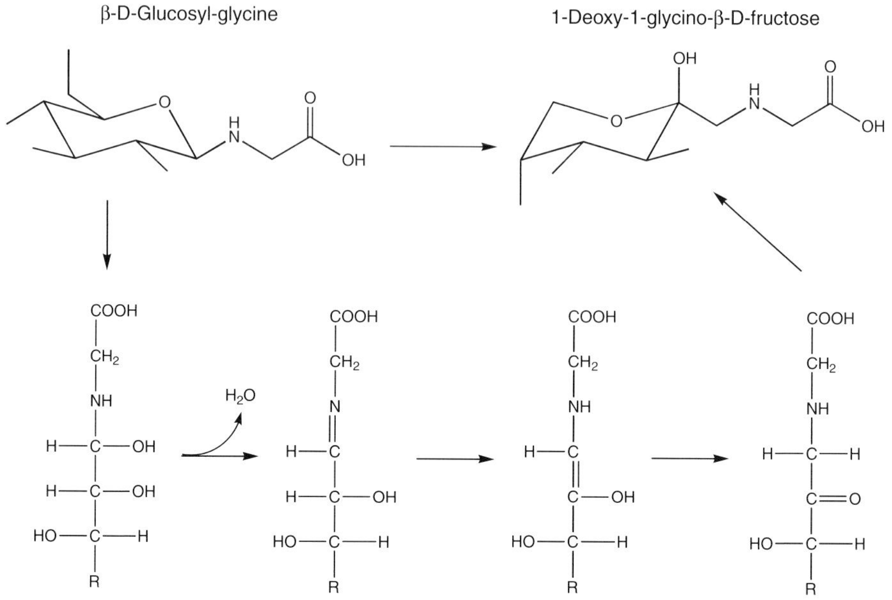

β-D-Glucosyl-glycine   1-Deoxy-1-glycino-β-D-fructose

Figure 4.2   The Amadori rearrangement: Formation of an amino-substituted ketose from an aldosyl-amine.

A ketosyl amine will undergo a similar rearrangement to form an amino-substituted aldose. The formation of 2-deoxy-2-glycino-β-D-glucose from β-D-fructosyl-glycine is shown in Figure 4.3.

Dehydration, Enolization, and Rearrangement Reactions

Under acidic conditions, the Amadori compound undergoes proton-assisted loss of a hydroxyl group at carbon-3, water elimination, 2-3-enolization, loss of the amino acid through hydrolysis, and enol-keto transformation to form a 3-deoxyosulose (Figure 4.4). Regeneration of the amino acid accounts for why only a small concentration of amino acids need be present for Maillard browning to occur.

Under more alkaline conditions, the Amadori compound undergoes base-assisted loss of a proton at carbon-3, loss of water, and loss of the amino compound through hydrolysis to generate the amino acid and 1-methyl-2-3-dicarbonyls (Figure 4.5). The Hodge scheme (Figure 4.1) shows conversion of the 1-methyl-2-3-dicarbonyl to an intermediate having two enolic hydroxyls at carbon-3 and carbon-4 in conjugation with the carbonyl at carbon-2. Compounds having two enolic hydroxyls in conjugation with a carbonyl are known as reductones. Ascorbic acid is a reductone.

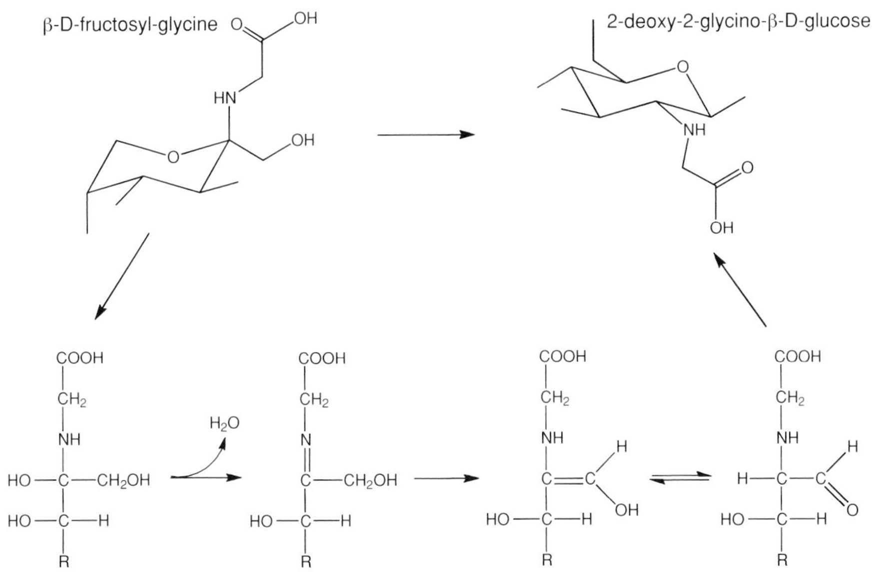

Figure 4.3 The Heyn's rearrangement: Formation of an amino-substituted aldose from a ketosyl-amine.

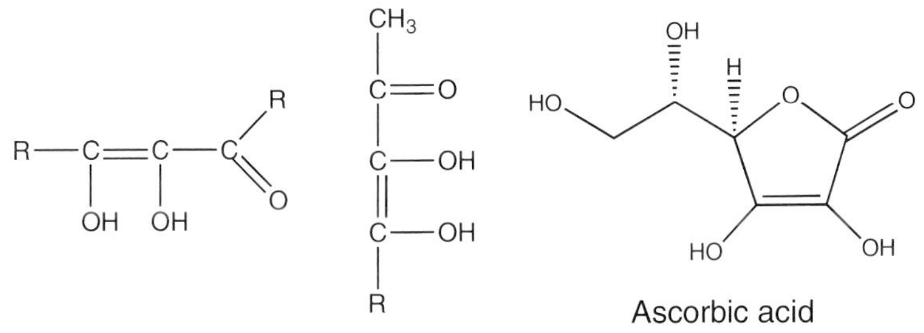

Ascorbic acid

Structure 4.4   Reductones.

Generation of reductones through Maillard browning can account for some of the apparent increases in antioxidant properties of some foods during processing and storage as measured by oxygen radical absorbance capacity (ORAC) and ferric reducing antioxidant power (FRAP) (Nursten 2005).

The Strecker Degradation

Referral to the Hodge scheme (Figure 4.1) shows the generation of α-dicarbonyls under both acidic and more alkaline conditions. α-Dicarbonyls are reactive compounds that will react with an amino acid

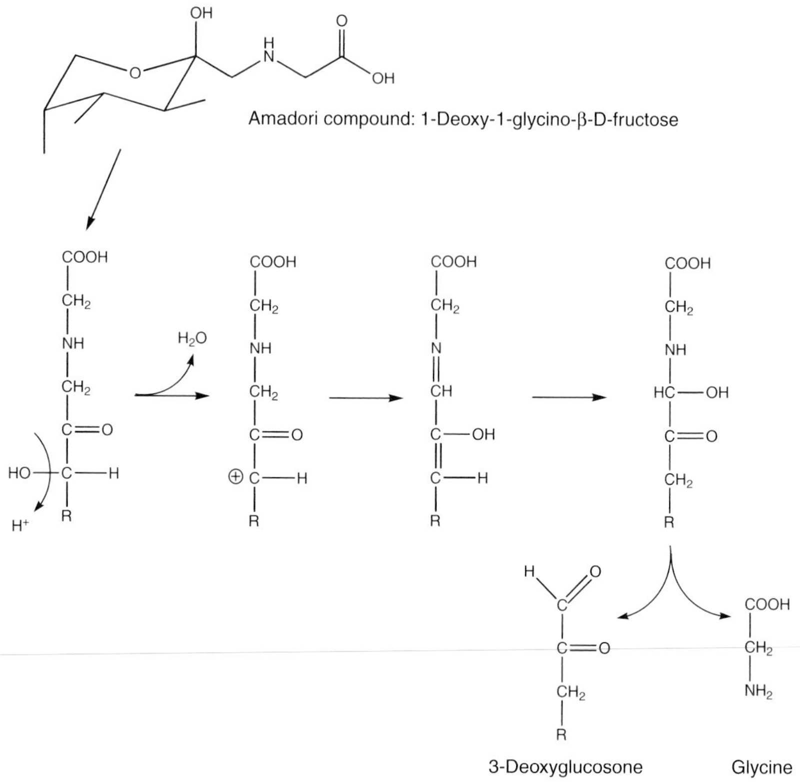

Amadori compound: 1-Deoxy-1-glycino-β-D-fructose

Figure 4.4 **Transformation of the Amadori compound to 3-deoxyosulose and an amino acid via 2-3-enolization.**

to generate CO_2 and an aldehyde with a chain length one carbon less than the amino acid. The nitrogen-containing fragment can rearrange and break down to generate ammonia that is transferred to other components of the system. Two fragments can condense to form various pyrazines, which are potent flavor compounds found in coffee, roasted nuts, and other foods.

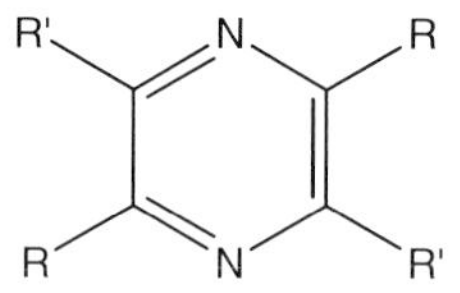

Structure 4.5   Pyrazines.

In addition to its flavor significance, the Strecker degradation (Figure 4.6) accounts for the formation of CO_2 during storage and the loss of amino acids and proteins.

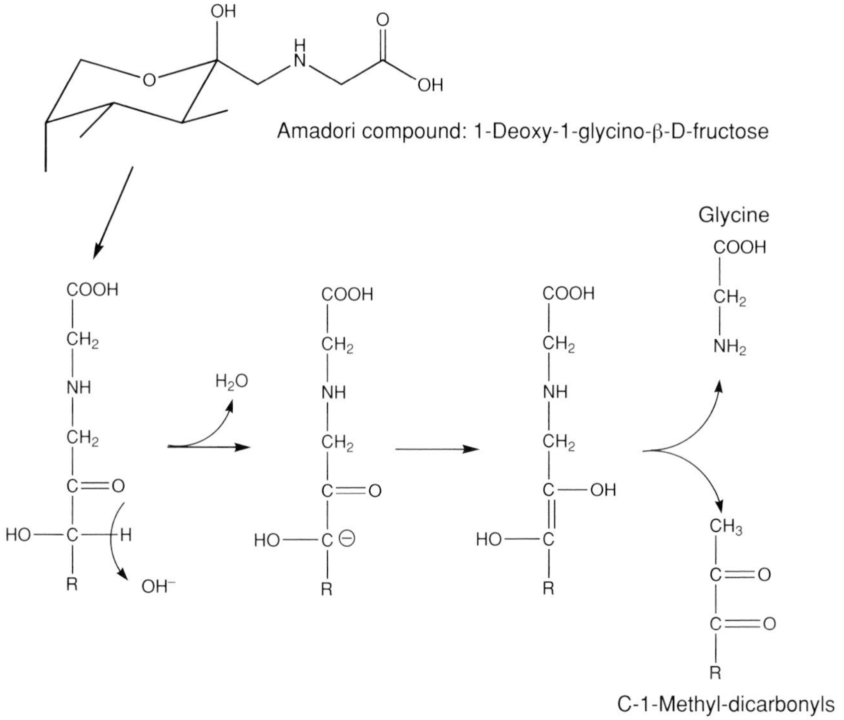

Figure 4.5   **Transformation of an Amadori compound to C-1-methyl-dicarbonyls and free amino acids.**

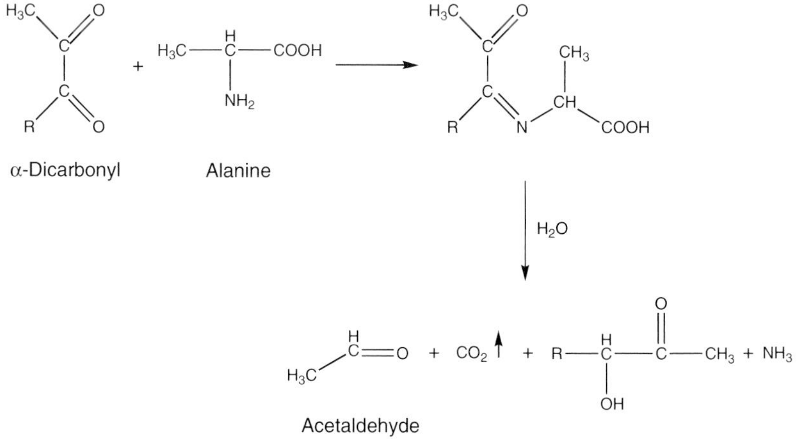

Figure 4.6   **The Strecker degradation: Reaction of α-dicarbonyls with amino acids.**

Final Stages: Condensation and Polymerization

The compounds formed in the Maillard browning scheme presented so far are colorless. The initial condensation product and the Amadori compounds are colorless and non-ultraviolet (UV) absorbers. The unsaturated carbonyls and reductones are strong UV-absorbers and exhibit some fluorescence, but they do not absorb in the visible spectrum. Polymerization of carbonyls through the aldol condensation is generally accepted as a primary route for formation of polymeric melanoidin pigments. Increased conjugation through dehydration will eventually lead to colored compounds. Concentration of reactants and dehydrating conditions will favor the reaction.

Reaction 4.6   The aldol condensation.

Some key features of the reactions discussed above are the numerous mechanisms that involve ring opening and the importance of conditions favoring dehydration and condensation. Amino acids can be regarded as catalysts because they are regenerated in the scheme, and only small amounts are required for the reactions to take place. Compounds such as furfural and 3-deoxyglucosone can be formed by acid-catalyzed dehydration of sugars (see Chapter 3). The Maillard reaction produces these same compounds but under milder reaction conditions. Amino acids through the complex series of reactions effectively lower the energy of activation.

An Alternate Free-Radical Mechanism for Nonenzymatic Browning

Namiki can be credited with the paradigm shift that considers an alternate mechanism for nonenzymatic browning involving free-radical formation (Namiki and Hayashi 1983; Namiki 1988). The presence of stable free radicals in roasted coffee and other browning reaction products had been detected by electron spin resonance (ESR) as early as 1953 (Rizzi 2003). Namiki and Hayashi established that stable free radicals were generated in the initial stages of the Maillard reaction. They detected the presence of N,N-dialkylpyrazine cation radicals, which were present prior to the

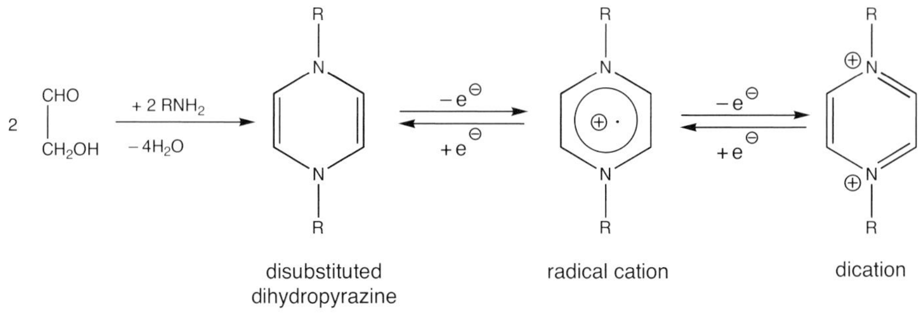

Figure 4.7   Scheme for formation of N,N-dialkylpyrazine cation radicals from 2-carbon sugar fragments.
Source: Nursten 2005, courtesy of Springer.

formation of Amadori compounds. They proposed that 2-carbon sugar fragments were precursors to the radical cation, which subsequently degrades to form melanoidin pigments (Figure 4.7).

It should be emphasized that Namiki's free radical mechanism does not displace the classical Hodge scheme. It is widely accepted that both routes occur (Eskin 1990, Rizzi 2003, Nursten 2005). Hayashi (1988) showed that formation of 2- and 3-carbon fragments were greatly influenced by pH with negligible formation at acidic pH, observable at neutral and substantial increases under alkaline conditions. Cämmerer and Kroh (1996) monitored the proportions of free radical and ionic intermediates and confirmed that pH had a profound impact on the involvement of free radical intermediates. At pH 5, the ionic mechanism predominated, with an increasing amount of radical participation with increasing pH.

Measurement of Maillard Browning

Absorbance at 420 nm, or a similar wavelength in the lower range of the visible spectrum, is most often used to measure Maillard browning of liquid foods and extracts. The absorbance spectrum of melanoidin pigments is very broad, so the precise wavelength is not critical as long as one is consistent. (A sharp absorption band at 420 nm would present a yellow color.) Tristimulus colorimeters are often used to monitor browning during storage, particularly for solid foods. L*, a measure of darkness and lightness with 100 = white and 0 = black, is the most useful parameter from CIEL*a*b* measurements. Occasionally, measurement of fluorescence and UV absorbance are used. During long-term storage of foods, CO_2 production from the Strecker degradation is sometimes measured. Disappearance of reactants (reducing sugars) is very rarely used to measure Maillard browning. The latter would give a measure of the initial condensation reaction, whereas the other measurements are of the end products of a series of complex reactions.

Control of Maillard Browning

Introductory Comments

What is important to food scientists is how the above knowledge can be applied to better control browning in foods during processing and storage. The following material is strongly influenced by the excellent chapter by Eskin (1990), which is highly recommended.

Water Activity

The degree of browning is heavily influenced by water activity (A_w). Figure 4.8 shows the effect of A_w on the degree of browning in an Avicel®-glucose-glycine model system after 8 days at 38°C (McWeeny 1973). At very low A_w, the browning reaction is virtually stopped because of immobilization of reactants. With increasing A_w, browning increases at a

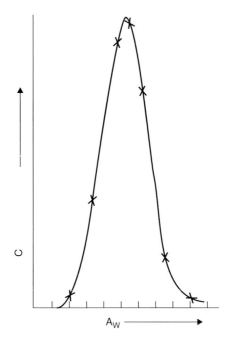

Figure 4.8   Browning of an avicel-glucose-glycine model system as affected by water activity after 8 days at 38°C.
Source: McWeeny 1973, with permission.

pronounced rate. Water serves as a solvent for the reaction medium, and water is also a reactant participant, being required for hydrolysis, enolization, and acid-base catalysis. With increasing A_w, browning slows because of dilution of reactants. Intermediate moisture foods ($A_w = 0.6$–0.9) are particularly susceptible to Maillard browning; hence, it is a major issue in jams and preserves, fruit juice concentrates, and dried fruits. Although these foods are microbiologically stable at room temperature, Maillard browning reactions will occur under such conditions.

The Importance of pH

The impact of pH on the amount of browning in a glucose-glycine model system is shown in Figure 4.9. The rate of browning increases dramatically as pH increases from 4 to the alkaline range. The condensation

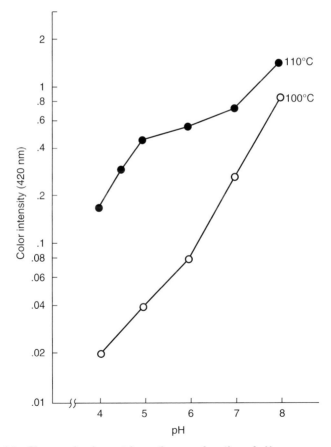

Figure 4.9   Changes in pigment formation as a function of pH.
Source: Lee et al. 1984, with permission.

reaction will be favored with an increase in the free-base form of the amino group, and 2-3-enolization of the Amadori compound to produce C-1-methyldicarbonyls and reductones will be favored. In addition, the free-radical route for melanoidin pigment formation becomes more pronounced with increasing pH. The relatively high pH of milk-based systems enables the development of the desired brown color and flavor compounds in caramel confections and dulce de leche. Pretzels are given an alkali dip to hasten formation of brown color. Alkali also gelatinizes the starch granules (see Chapter 7) at the surface, which facilitates light transmission that contributes to the attractive visual appearance of pretzels.

Nature of Reactants

Replacement of reducing sugars with nonreducing sugars is one approach to preventing Maillard browning. In surimi manufacture, carbohydrates serve as **cryoprotectants** in the formulation. The protein gel structure is protected from ice crystal damage by rendering the water molecules unfreezable by partial immobilization. Surimi is used in producing artificial crab and other seafood analogues, and whiteness is a critical quality standard. To prevent the browning reaction, nonreducing sugars, such as trehalose and sucrose, and sugar alcohols (e.g., sorbitol and reduced maltodextrins) are used. These sugars and sugar derivatives vary considerably in sweetness, which makes it possible to obtain the desired level of sweetness in the surimi product.

Egg white is one of the few alkaline foods (pH $= 7.6–7.9$). The presence of significant quantities of glucose in egg white results in browning and loss of protein functionality during the spray-drying process because of the Maillard reaction. Removal of glucose by treating egg white with glucose oxidase and catalase effectively removes glucose to give a more acceptable product.

Reaction 4.7 Removal of glucose from egg white by treatment with glucosse oxidase and catalase.

Although glucose is overwhelmingly the most studied sugar in Maillard browning experiments, it is the least reactive of the hexose and pentose sugars (Table 4.1). Pentose sugars are more reactive than hexoses, with ribose being the most reactive because it has such a high proportion in the open-chain form.

Although amino acids are considered catalysts in the Maillard reaction scheme, the quantity of amino acids can still affect reaction rate. Pear juice concentrate has a greater propensity for browning than apple juice concentrate, even though their sugar profile, pH, and water activity are similar. The larger quantity of total amino acids in pear juice is believed to account for its greater rate of browning.

Case Study

Pear Juice Concentrate: Removal of amino acids from pear juice by treatment with cation exchange resin gave a product with a similar sugar and nonvolatile acid profile and pH to the untreated juice (Cornwell and Wrolstad 1981). Figure 4.10 compares the rate of browning during storage at 37°C for the control and concentrates treated with Polyclar AT™ (PVPP), Amberlite XAD-4 resin, and Dowex AG 50 cation exchange resin. Although

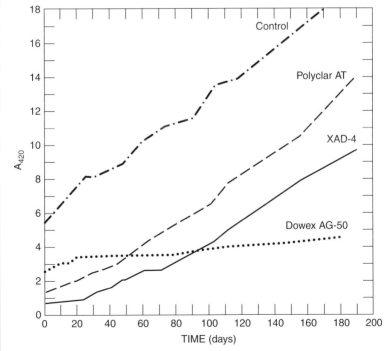

Figure 4.10   Increase in browning for pear juice concentrate (control) and concentrates treated with Polyclar AT™, Amberlite XAD-4 resin, and Dowex AG-50 cation exchange resin.
Source: **Cornwell and Wrolstad 1981.**

PVPP and XAD-4 resins were most effective in reducing initial color, treatment with cation exchange resin effectively stopped browning by removing the amino acids. Although the color and taste properties of the resin-treated juice make it a desirable beverage and canned fruit ingredient, its use cannot be legally labeled as pear juice. Removal of amino acids and altering the mineral profile render it "decharacterized juice."

Case Study

Cantaloupe Juice Concentrate: With the objective of using surplus fresh market cantaloupes and sort-outs, a process was developed for producing

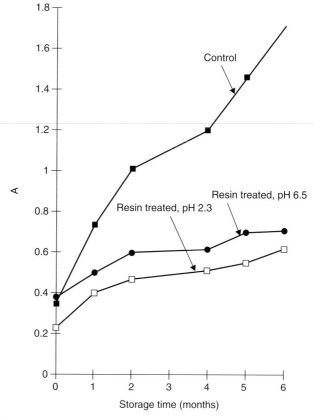

Figure 4.11   Browning of clarified cantaloupe juice concentrate (pH 6.5), concentrate treated with Dowex cation exchange resin (pH 2.3), and concentrate treated with Dowex cation exchange resin with pH adjusted to 6.5 during storage at 25°C for 6 months.
Source: Galeb 1993.

decharacterized cantaloupe juice that could serve as a sweetener (Galeb et al. 2002). The relatively high amino acid and ascorbic acid content, along with a pH of 6.5, makes cantaloupe juice concentrate particularly susceptible to browning. Figure 4.11 shows the rate of browning for the control cantaloupe juice concentrate and cantaloupe juice treated with cation exchange resin. Because ion-exchange treatment lowered the pH to 2.3, the pH of one lot of the treated juice was adjusted to 6.5 in order to examine the effect of both amino acid removal and pH. Removal of amino acids with cation exchange markedly reduced the rate of browning (Figure 4.11). Browning that occurred during juice concentration (value at time zero) was lower at pH 2.3. The rates during storage of the resin-treated concentrates, however, were not significantly different.

Although there have been many investigations on the effect of sugar composition on Maillard browning, less attention has been paid to the impact of amino acid composition. Ashoor and Zent (1984) investigated the degree of browning of individual amino acids in glucose, fructose, and ribose model systems and sorted them into three groups of reactivity: high, intermediate, and low, as summarized in Table 4.2.

In April 2002, the Swedish National Food Authority reported the presence of elevated levels of acrylamide in certain types of food that had been processed at high temperatures (Taeymans et al. 2004). This caused immediate concern because acrylamide is known to cause cancer in laboratory animals. Investigators from many countries soon found trace amounts of acrylamide in a wide variety of foods, including potato chips, French fries, breads, and other baked cereal goods. Its formation (Figure 4.12) was caused by the reaction of asparagine with reducing sugars under high temperature and low-moisture conditions (Zhang and Zhang 2007). Efforts have been made to reduce or eliminate its formation by altering process conditions. One innovative process is to treat dough with asparaginase (Acrylaway®) to reduce the concentration of asparagine; this has been successfully applied to crisp breads, crackers, and snack foods. Another approach is to select plant genotypes with reduced amounts of asparagine.

Temperature

Reducing storage temperature is often the only recourse one has for reducing Maillard browning. The rate of browning will be reduced by two to three times with a $10°C$ lowering of temperature. Maillard browning is essentially negligible at $-20°C$. The case study of shipment of Argentine apple juice concentrate described below illustrates this application.

Table 4.2   Comparative Browning Intensity of Amino Acids in Sugar Model Systems[a]

L-Amino Acid	Absorbance at 420 nm		
	D-Glucose	D-Fructose	D-Ribose
High Reactivity			
Lysine	0.947	1.04	1.22
Glycine	0.942	1.07	1.34
Tryptophan	0.826	0.853	0.972
Tyrosine	0.809	0.857	0.951
Intermediate Reactivity			
Proline	0.770	0.783	0.792
Leucine	0.764	0.747	0.895
Isoleucine	0.746	0.797	0.870
Alanine	0.739	0.792	0.945
Hydroxyproline	0.738	0.752	0.813
Phenylalanine	0.703	0.751	0.800
Methionine	0.668	0.669	0.828
Valine	0.663	0.800	0.772
Glutamine	0.602	0.644	0.633
Serine	0.600	0.646	0.679
Asparagine	0.560	0.578	0.565
Low Reactivity			
Histidine	0.535	0.573	0.529
Threonine	0.509	0.601	0.590
Aspartic acid	0.353	0.426	0.378
Arginine	0.335	0.331	0.370
Glutamic acid	0.294	0.338	0.341
Cysteine	0.144	0.202	0.150

[a]Amino compound:sugar molar ratio of 1:1, final concentration of 0.05M in 0.04M carbonate buffer of pH 9.0. The solutions were autoclaved at 121°C for 10 min.
Source: Modified from Ashoor and Zent 1984, with permission.

Case Study

Shipment of Argentine Apple Juice Concentrate: It used to be common practice to ship fruit juice concentrates at ambient temperature over long distances. Our laboratory participated in a collaborative project with Argentine producers of apple juice concentrate to determine the extent of nonenzymatic browning and quality deterioration that occurred from shipment of Argentine apple juice concentrate across the equator to the United States (Babsky et al. 1989). Nine lots were sampled at a processing plant in Argentina, and the same lots were sampled again after 53–55 days transit to Modesto, California. The magnitude of change (Table 4.3) was sufficient for Argentine producers to change to refrigerated shipment, which is commonplace today.

Table 4.3   Browning Indices[a] of Argentine Apple Juice Concentrate Before and After Shipment to the United States[b]

	A_{420nm}	$\%T_{625nm}$	HMF	L	$(a^2 + b^2)^{1/2}$
		Before Shipment			
Range	0.462–0.534	92–94	5.6–8.1	88.8–89.4	25.9–28.8
Mean	0.504	93	6.9	89.1	27.3
		After Shipment			
Range	0.593–0.688	88–91	7.5–12.5	85.8–88.3	30.5–34.9
Mean	0.623	89	9.8	87.2	32.6

[a]A_{420nm} = absorbance at 420 nm; $\%_{625nm}$ = % Transmittance at 625 nm; HMF = hydroxymethylfurfural, mg/L; L = Hunter L value; $(a^2 + b^2)^{1/2}$ = chroma.
[b]Modified from Babsky et al. 1989, with permission. Differences for all parameters were significant at the 99% confidence level.

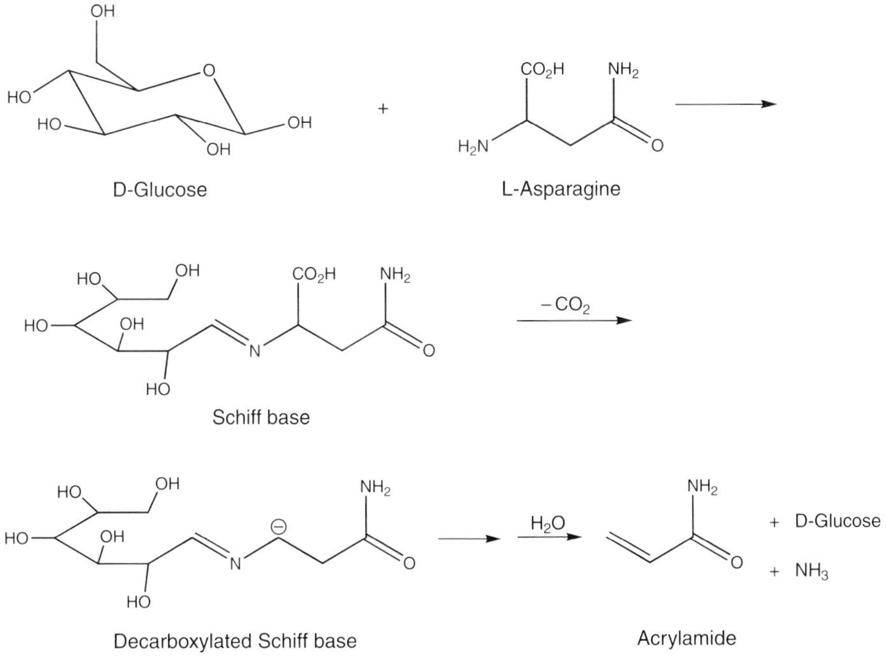

Figure 4.12   Proposed mechanism for formation of acrylamide from the reaction of arginine with glucose.
Source: Bemiller and Huber 2008.

Oxygen

None of the reaction schemes in this chapter shows oxygen as a reactant in Maillard browning. Oxygen is not a requirement, and removal of headspace oxygen will not prevent the reaction from occurring. Oxidation of reactive intermediates can alter reaction routes, but it could conceivably decrease rather than increase the degree of browning. (Oxidation of reactive aldehydes to carboxylic acids could reduce the amount of polymer formation.)

Chemical Inhibitors

McWeeny investigated the use of SO_2 for control of Maillard browning (McWeeny 1973). His hypothesis was that reactive carbonyls would form sulfonic acid adducts with bisulfite, and thus be unavailable for polymerization and melanoidin pigment formation. Although he found that SO_2 could delay the reaction to a small extent, its use was not practical for industrial application.

Reaction 4.8   Reaction of aldehydes with SO_2 (McWeeny 1973).

Other Browning Reactions

Caramelization

Sugar caramelization is defined as the production of characteristic flavors and brown pigments from sugars by heat application in absence of amino compounds. By thermolysis, sugars are dehydrated by 1-2- and 2-3-enolization, dependent on pH, and then fragmented to reactive compounds that polymerize to colloidal brown polymers and also form typical flavors. Ironically, milk-based caramel confections do not meet the above definition of caramelization.

Caramel coloring is an approved color additive that is exempt from certification. It is produced by controlled heating of sugars at varying pHs. Some caramel colorings are manufactured without the use of ammonium catalysts and thus are in keeping with the above definition of carmelization. However, ammonium salts are used in many of the product lines, and Amadori compounds are formed. Colorant companies produce a variety of caramel colorings with different hues and varying colloidal charge and size for different food applications. Caramel coloring has the largest usage of all food colorants worldwide (Sepe et al. 2008).

Ascorbic Acid Browning

Ascorbic acid is a reductone, and it will form brown pigments in the presence or absence of oxygen and in the presence or absence of amino compounds. Figure 4.13 shows the oxidative degradation of ascorbic acid to give 2,3-diketogulonic acid, furfural, and CO_2. This is a somewhat simplified scheme as the reactions are much more complex than that presented. A common misconception is that ascorbic acid will prevent Maillard browning. That is not the case; rather, ascorbic acid will accelerate Maillard browning.

Enzymatic Browning

Although not a reaction of sugars, enzymatic browning merits mentioning because the methods for its control contrast with those for nonenzymatic browning, and the two are often confused. Enzymatic browning occurs when native polyphenol oxidase in the presence of oxygen oxidizes

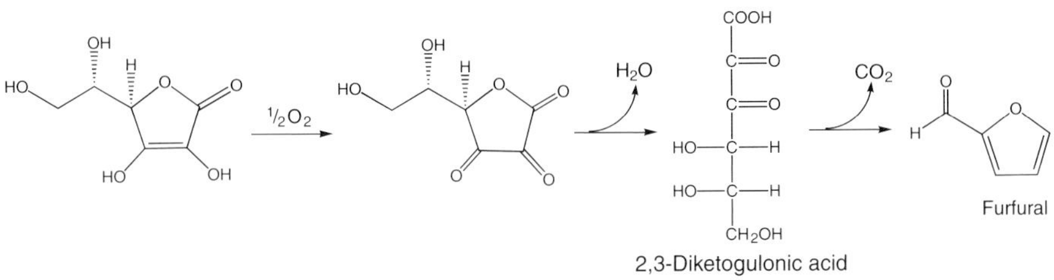

Figure 4.13 Oxidative degradation of ascorbic acid to form 2,3-diketogulonic acid, furfural, and CO_2.

o-phenolic compounds to quinones, which subsequently polymerize to form brown pigments.

Reaction 4.9   Scheme for polyphenol oxidase catalyzed formation of melanoidin pigments.

Oxygen is a requirement for the reaction to occur, and bisulfite is an effective inhibitor of the enzyme. Ascorbic acid will reduce the quinones to phenols and reverse the reaction.

Assessing Contributing Factors to Nonenzymatic Browning

There is practical interest in understanding the relative contributions of the Maillard reaction, caramelization, and ascorbic acid browning in food systems.

Case Study

Modeling Browning of Kiwi Fruit Juice Concentrate: Kiwi fruit juice suffers from pronounced browning, which is partially attributed to its high ascorbic acid content. Wong and Stanton (1989) performed a modeling experiment in which they used cocktails of sugars and organic acids having the same profile as kiwi fruit, ascorbic acid, and a combination of amino acids to investigate the relative contributions of different types of browning. Sugars and acids combined with amino acids were defined as the Maillard browning contribution. Sugars and acids in absence of amino compounds were defined as caramelization. Sorbitol in combination with ascorbic acid and amino acids was defined as Strecker degradation, and sorbitol and ascorbic acid was defined as ascorbic acid browning. Sugars, acids, amino acids, and ascorbic acid in combination represented overall browning. Their results are shown in Table 4.4. In decreasing order, the browning rates were ascorbic browning > Maillard browning > Strecker browning > caramelization. When the rates of the individual browning reactions are summed, they are equivalent to the rate for the overall system, suggesting that the browning reactions are independent and not synergistic. The browning rate for the overall system was considerably lower than that measured for actual kiwi fruit juice concentrate. A plausible explanation is that, in the case of kiwi fruit juice concentrate, reactive intermediates had been formed in juice concentrate production. This would not have been the case for the model.

Table 4.4   Browning Rates for Models[a] of Kiwi Fruit Juice Concentrate[b]

Model Composition	Browning Reaction Model	Nonenzymatic Browning Rate[c] A_{420nm}.week^{-1}
Sugars + Nonvolatile Acids	C	0.0007
Sugars + Nonvolatile Acids + Ascorbic Acid	C + A	0.0048
Sugars + Nonvolatile Acids + Amino Acids	C + M	0.0042
Sorbitol + Nonvolatile Acids + Amino Acids + Ascorbic Acid	A + S	0.0030
Sugars + Nonvolatile Acids + Amino Acids + Ascorbic Acid	C + M + A + S	0.10

[a]C = caramelization; A = ascorbic acid browning; M = Maillard browning; S = Strecker degradation.
[b]Modified from Wong and Stanton 1989, with permission.
[c]Rates measured for storage at 20°C for 10 weeks.

Case Study

Modeling Potato Chip Browning (Table 4.5): The characteristic brown color of potato chips is an important quality attribute that is caused by

Table 4.5   Modeling Potato Chip Browning[a]

Model Composition[c]	CIEL*a*b* Color Indices[b]		
	L*	Chroma	Hue Angle
Complete set	46.5	39.8	68.5
Complete set minus Sucrose	45.5	39.0	68.6
Complete set minus Ascorbic	49.1	38.6	70.1
Complete set minus Amino acids	49.3	33.8[d]	75.6[d]
Complete set minus Reducing sugars	57.2[d]	31.3[d]	86.0[d]

[a]Modified from Rodriguez-Saona et al. 1997. Leached potato slices were infused with model solutions and fried for 3 min at 180°C.
[b]L* = Darkness/Lightness (0 = black; 100 = white); Chroma = $(a^{*2} + b^{*2})^{1/2}$; Hue angle = arctan b*/a*.
[c]Complete set = reducing sugars (glucose + fructose) + sucrose + ascorbic acid + amino acids (L-glutamine and L-asparagine).
[d]Indicates significant difference among means ($P < 0.01$).

Maillard browning. Reducing sugar levels vary with storage conditions and with cultivar, as does ascorbic acid and amino acid content. In an effort to determine the relative contribution of sugars, ascorbic acid, and amino acids, Rodriguez-Saona et al. (1997) took an approach similar to that of Wong and Stanton (1989). Slices of potato were washed with aqueous ethanol to remove all sugars, ascorbic acid, and amino acids. (This material when deep fat fried had very little color.) The slices were infused with solutions having component concentrations typical of potatoes. The slices were fried for 3 min at 180 C, and the color attributes measured by CIEL*a*b*. The complete solution set consisted of reducing sugars (glucose and fructose), sucrose, ascorbic acid, and amino acids (L-glutamine and L-asparagine). Elimination of reducing sugars had the greatest impact on browning as evidenced by L* and chroma.

Vocabulary

Caramelization—The production of characteristic flavors and brown pigments from sugars by heat application in absence of amino compounds

Cryoprotectant—Compound that protects foods and biological tissues from freeze damage due to ice crystal formation

Maillard browning—The formation of melanoidin pigments and characteristic flavor compounds in a complex series of reactions initiated by the reaction of reducing sugars and amino compounds

Strecker degradation—Reaction of α-dicarbonyls with amino acids to generate CO_2 and an aldehyde with a chain length one carbon less than the amino acid.

References

Ames JM. 1998. Applications of the Maillard reaction in the food industry. *Food Chem* 62:431–439.

Ashoor SH, Zent JB. 1984. Maillard browning of common amino acids and sugars. *J Food Sci* 49:1206–1207.

Babsky NE, Wrolstad RE, Durst RW. 1989. Influence of shipping on the color and composition of apple juice concentrate. *J Food Qual* 12:355–367.

Baynes JW, Monnier VM, Ames JM, Thorpe SR, editors. 2005. *The Maillard reaction. Chemistry at the interface of nutrition, aging, and disease.* Hoboken, NJ: Wiley-Blackwell.

BeMiller JN, Huber KC. 2008. Carbohydrates. In: Damodaran S, Parkin KL, Fennema OR, editors. *Fennema's food chemistry, 4th ed.* New York: CRC Press, pp. 83–154.

Burton HS, McWeeny DJ. 1963. Non-enzymatic browning reactions: consideration of sugar stability. *Nature* 197:266–268.

Cämmerer B, Kroh LW. 1996. Investigation of the contribution of radicals to the mechanism of the early stage of the Maillard reaction. *Food Chem* 57:217–221.

Cornwell CJ, Wrolstad RE. 1981. Causes of browning in pear juice concentrate during storage. *J Food Sci* 46:515–518.

Eskin NAM. 1990. Biochemistry of food processing: browning reactions in foods. In: Eskin NAM. *Food biochemistry, 2nd ed.* New York: Academic Press, pp. 239–296. *[Note: This chapter was required reading in the Food Chemistry and Food Carbohydrates classes that I taught. It is highly recommended.]*

Galeb ADS. 1993. Use of ion-exchange and direct osmotic concentration technologies for processing cantaloupe juice. Doctoral thesis. Corvallis, OR: Oregon State University.

Galeb ADS, Wrolstad RE, McDaniel MR. 2002. Composition and quality of clarified cantaloupe juice concentrate. *J Food Proc Pres* 26:39–56.

Gerrard JA. 2006. The Maillard reaction in food: progress made, challenges ahead. Conference report from the 8th International Symposium on the Maillard reaction. *Trends Food Sci Technol* 17:324–330.

Hashiba H. 1982. The browning reaction of Amadori compounds derived from various sugars. *Agr Biol Chem* 46:547–548.

Hodge JE. 1953. Dehydrated foods: chemistry of browning reactions in model systems. *J Agric Food Chem* 1:928–943. *[Note: This Citation Classic is a review that, for the first time, provided a comprehensive organization of the reactions involved in the production of brown pigments in foods heated to moderate temperatures. It is the most highly cited article from J Agric Food Chem as of 2009.]*

Hodge JE. 1967. Origin of flavor in foods: nonenzymatic browning reactions. In: Schultz HW, Day EA, Libbey LM, editors. *Symposium on foods: the chemistry and physiology of flavors.* Westport, CT: AVI Publishing, pp. 465–491.

[IMARS] International Maillard Reaction Society. 2010. International Maillard Reaction Society website. Available from: http://www.imars.org/online/. Accessed September 23, 2010.

Labuza TP, Schmidl MK. 1986. Control of browning reactions in foods. In: Fennema OR, Chang W-H, Lii C-Y, editors. *Role of chemistry in the quality of processed food.* Westport, CT: Food & Nutrition Press, pp. 65–95.

Lee CM, Sherr B, Koh Y-N. 1984. Evaluation of kinetic parameters for a glucose-lysine Maillard reaction. *J Agric Food Chem* 32:379–382.

Maillard LC. 1912. Action des acides amines sur les sucres: formation des melanoidines par voie methodique. *C R Hebd Seances Acad Sci* 154:66–68.

Manzocco L, Calligaris S, Mastrocola D, Nicoli MC, Lerici CR. 2001. Review of non-enzymatic browning and antioxidant capacity in processed foods. *Trends Food Sci Technol* 11:340–346.

Martins SIFS, Jongen WMF, van Boekel MAJS. 2000. A review of Maillard reaction in food and implications to kinetic modeling. *Trends Food Sci Technol* 11:364–373.

McWeeny DJ. 1973. The role of carbohydrate in non-enzymatic browning. In: Birch GG, Green LF, editors. *Molecular structure and function of food carbohydrate*. New York: John Wiley & Sons, pp. 21–32.

Namiki M. 1988. Chemistry of Maillard reactions: recent studies on the browning reaction mechanism and the development of antioxidants and mutagens. In: Chichester CO, Schweigert BS, editors. *Advances in food research, vol. 32*. New York: Academic Press, pp. 115–184.

Namiki M, Hayashi T. 1983. A new mechanism of the Maillard reaction involving sugar fragmentation and free radical formation. In: Waller GR, Feather MS, editors. *The Maillard reaction in foods and nutrition*. ACS Symposium Series 215. Washington, D.C.: American Chemical Society, pp. 21–46.

Nursten HE. 2005. *The Maillard reaction: chemistry, biochemistry, and implications*. Cambridge, UK: The Royal Society of Chemistry. *[Note: This single-author monograph is a concise and comprehensive survey of the Maillard reaction and its role in food science. The author has particularly made significant contributions to the reactions role in flavor generation.]*

O'Brien J, Nursten HE, Crabbe JC, Ames JM. 1998. *The Maillard reaction in foods and medicine*. Cambridge, UK: The Royal Society of Chemistry.

Rizzi GP. 2003. Free radicals in the Maillard reaction. *Food Rev Intern* 19:375–395.

Rodriguez-Saona LE, Wrolstad RE, Pereira C. 1997. Modeling the contribution of sugars, ascorbic acid, chlorogenic acid and amino acids to non-enzymatic browning of potato chips. *J Food Sci* 62:1001–1005.

[SCI] Science Citation Index. 2010. Science Citation Index. Available from: http://thomsonreuters.com. Accessed September 20, 2010.

Sepe HA, Parker OD, Nixon AR, Kamuf WE. 2008. Global color quality of beverages utilizing caramel color. In: Culver CA, Wrolstad RE, editors. *Color quality of fresh and processed foods*. ACS Symposium Series 983. Washington, D.C.: American Chemical Society, pp. 226–240.

Shellhammer TH, Bamforth CW. 2008. Assessing color quality of beer. In: Culver CA, Wrolstad RE, editors. *Color quality of fresh and processed foods*. ACS Symposium Series 983. Washington, D.C.: American Chemical Society, pp. 192–202.

Somoza V. 2005. Five years of research on health risks and benefits of Maillard reaction products: an update. *Mol Nutr Food Res* 49:663–672.

Taeymans KD, Wood J, Ashby P, Blank I, Studer A, Stadler RH, Gondé P, Van Eijk P, Lalljie S, Lingnert H, Lindblom M, Matissek R, Müller D, Tallmadge D, O'Brien J, Thompson S, Silvani D, Whitmore T. 2004. A review of acrylamide: an industry perspective on research, analysis, formation, and control. *Crit Rev Food Sci Nutr* 44:323–347.

Waller GR, Feather MS, editors. 1983. *The Maillard reaction in foods and nutrition.* Washington, D.C.: American Chemical Society.

Wong M, Stanton DW. 1989. Nonenzymic browning in kiwifruit juice concentrate systems during storage. *J Food Sci* 54:669–673.

Zhang Y, Zhang Y. 2007. Formation and reduction of acrylamide in Maillard reaction: a review based on the current state of knowledge. *Crit Rev Food Sci Nutr* 47:521–542.

5 Functional Properties of Sugars

Introduction

Sugars are multifunctional. In addition to their chemical, biological, and nutritional properties, their different taste and physical characteristics render them suitable for a wide variety of food applications. Understanding the underlying scientific principles of these properties can help one to predict their impact on a food system and to select suitable methods for measuring their effects. Technical bulletins provided by ingredient suppliers, although designed to promote the advantages of their product line, nevertheless give very useful information for the product development scientist. Some basic knowledge helps to evaluate that information and make appropriate choices in selecting ingredients for desired applications.

Food Carbohydrate Chemistry, First Edition. Ronald E. Wrolstad.
© 2012 John Wiley & Sons, Inc. Published 2012 by John Wiley & Sons, Inc.

Taste Properties of Sugars

Sweetness is the immediate property that springs to mind when the word "sugar" is mentioned, and it is a quality characteristic of most sugars. Table 5.1 lists relative sweetness values for several selected sugars, sugar alcohols, and artificial sweeteners taken from the literature. The reference value for sweetness is typically taken as sucrose = 100. Relative sweetness values are useful, but their limitations need to be taken into consideration. First of all, the sensation of sweetness is not the same for all sugars. Figure 5.1 shows a time/intensity plot for the human response to fructose, sucrose, and glucose. The time for peak sweetness intensity to occur as well as the level of intensity is quite variable. Some sugars (e.g., mannose) have a substantial bitter component, which confounds assessment of sweetness.

Table 5.1 Relative Sweetness Values of Selected Sugars, Sugar Alcohols, and High-Intensity Sweeteners

Compound	Relative Sweetness (Sucrose = 100)		Source
	Range	Mean	
Sugars			
D-Fructose	100–180	121	Shallenberger 1993
D-Glucose	50–82	64	"
L-Glucose		64	"
D-Galactose	32–67	50	"
α-D-Mannose	32–59		"
β-D-Mannose	"Bitter"		"
Tagatose		92	Skytte 2006
Maltose	32–60	43	Shallenberger 1993
Lactose	16–45	33	"
Trehalose		50	Lindley 2006
Sugar Alcohols			
Xylitol		90	Daniel et al. 2007
Sorbitol		63	"
Galactitol		58	"
Maltitol		68	"
Lactitol		35	"
High-Intensity Sweeteners (Sucrose = 1)			
Saccharin	300–600		Nelson 2000
Cyclamate		30	"
Aspartame	160–220		"
Alitame		2000	"
Acesulfame K		200	"
Sucralose	400–800		"
Thaumatin	2000–3000		"
Glycyrrhizin	50–100		"
Stevioside	200–300		"

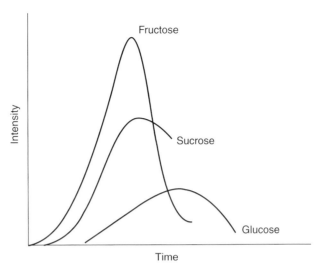

Figure 5.1   Partial sweetness taste profiles for common sugars. Modified from Shallenberger 1998, with permission.

Sugars at low concentrations (0.5–1.0%) will function as a flavor enhancer and, with increasing concentration, will be perceived as sweet. Soft drinks have a sugar concentration of approximately 10%, and confections are in the order of 30% or greater. Figure 5.2 shows the effect of increasing sucrose concentration and sweetness perception. Sweetness intensity is close to linearity from 1% to 27%, after which increasing concentration has reduced impact on sweetness perception. Other sugars will have a similar sigmoid shape but not identical to sucrose. Regarding the relative sweetness values for sugars in Table 5.1, different investigators may have used different sensory methods and varying concentration levels of sucrose as a reference. This could contribute to the considerable range in values for individual sugars. Temperature is another factor that can affect sweetness perception. Figure 3.2 (Chapter 3) shows the impact of solution temperatures from 5°C to 60°C on sweetness intensity of fructose, glucose, galactose, and maltose solutions. The marked difference in sweetness for fructose at different temperatures is explained by mutarotation equilibria, with higher amounts of the intensely sweet β-D-pyranose form being present at low temperatures.

In foods and beverages, sugars can interact with a wide variety of chemical constituents. Combinations of sugars can act synergistically in regard to sensory response. Some flavor compounds (e.g., maltol) can enhance sweetness. Bitter compounds can repress sweetness, and acids are very effective at repressing sweetness. Carbonic acid in soft drinks markedly reduces sweetness, as anyone who has drunk a "flat" soft drink will agree.

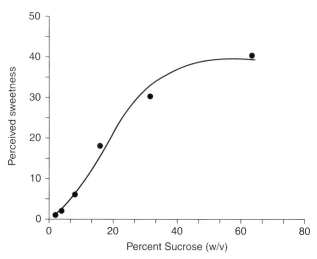

Figure 5.2 Effect of increasing concentration on the perceived sweetness of sucrose.
Source: **Shallenberger 1993.**

The Brix:Acid ratio (°Brix:Titratable Acidy) is widely used in monitoring and standardizing the taste quality of beverages. High-molecular-weight polysaccharides and proteins can complex with sugars and alter their perception. Because of these complex interactions, when making sweetener choices in product development, it is critical that the sweeteners be evaluated in the actual food system in which they will exist.

The Shallenberger–Acree Theory for Sweetness Perception

> **Anecdote**
>
> Bob Shallenberger gave a lecture on sugar stereochemistry and its relationship to sweetness perception while I was a graduate student at the University of California at Davis in 1963. Hydrogen bonding of vicinal OH groups to a receptor site and competition between intermolecular and intramolecular hydrogen bonding were part of the hypothesis. The audience was less than polite, highly critical of his hypothesis, and insistent that his data were insufficient. I had the good fortune to work with Bob Shallenberger in 1979–1980 when I was on sabbatical leave from Oregon State University at Cornell. One day I mentioned that I had attended his seminar in 1963. He smiled and said that he remembered the occasion very well, for it was the first time that he had presented his theory to a scientific audience.

He also stressed that scientific advances in his theory had been made by carefully listening to his critics, and then designing experiments to affirm or refute their criticisms.

Terry Acree joined Bob Shallenberger's laboratory as a graduate student in 1963, and they continued exploring why some compounds were sweet and others not sweet with the aid of molecular models and measurements of bond distances. In 1967, their article "Molecular Theory of Sweet Taste" was published in *Nature*, which has become a citation classic (Shallenberger and Acree 1967). Their theory proposed that sweetness perception was a concerted intermolecular hydrogen bonding interaction between the AH-B unit of a sweet-tasting compound and a complementary AH-B unit at the taste bud receptor site.

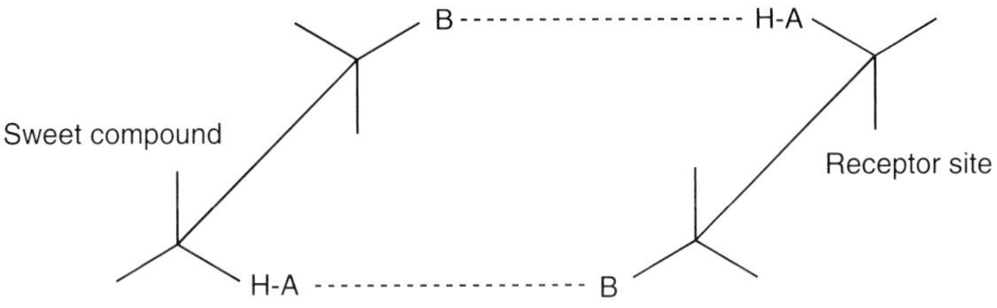

Structure 5.1

Specific stereochemical requirements are that the AH and B units be separated by approximately 3 angstroms. AH is a proton donor, and B a proton acceptor. An earlier observation (Shallenberger 1963) was that the ethylene glycol unit was the minimum structural unit for sweetness since ethylene glycol is sweet, and ethanol is not. (The sweetness of ethylene glycol accounts for the deaths of a number of dogs annually who lap up antifreeze to fulfill their passion for sweetness.) To elicit sweetness, the vicinal OH groups should have the *guache* or *staggered* conformation, and not be *eclipsed* or *anticlinal* (Shallenberger and Acree 1967). The adjacent OH groups in β-D-fructopyranose are gauche, whereas in β-D-fructofuranose they are eclipsed because of the nearly planar furanose ring.

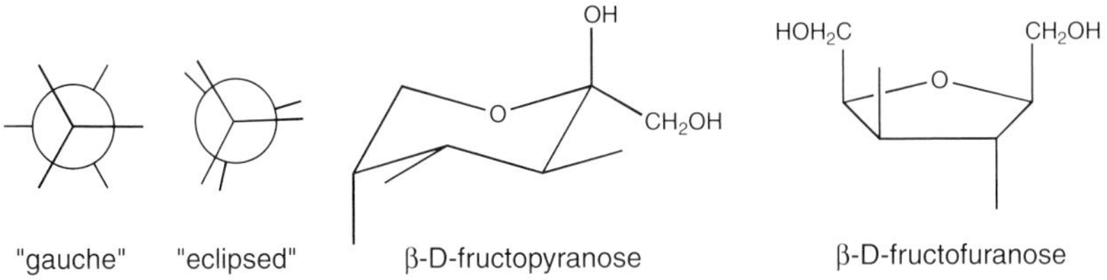

"gauche"   "eclipsed"   β-D-fructopyranose   β-D-fructofuranose

Structure 5.2

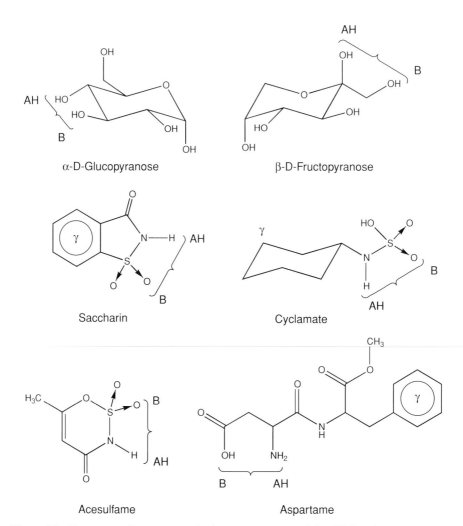

Figure 5.3   Representative compounds that taste sweet and the AH-B unit common to all of them. Modified from Shallenberger and Birch 1975.

The structures of glucose, fructose, and some intense sweeteners are shown in Figure 5.3, along with the designated AH-B units that meet the stereochemical requirement. A common feature of the intense sweeteners shown is that they are bipolar, containing a nonpolar moiety such as a benzene or cyclohexane ring. A hydrophobic or "greasy" component became an additional structural feature that could be attracted to a lipophilic component of the taste receptor (Shallenberger and Acree 1969).

Proof of the molecular site for sweetness is a challenging task. Birch and Lee (1974) performed some elegant experiments on the taste properties of a series of synthesized deoxy sugars. (These experiments would unlikely receive approval of a human subjects review panel today.) They demonstrated that the AH-B system for sugars consisted of the third and fourth carbon atom OH substituents, confirming Shallenberger and Acree's original proposal. Many investigators have continued to supplement the original AH-B model of Shallenberger and Acree (Acree and Lindley 2008). Much of the more recent work has been to better understand the structure and physiology of the sweetness receptor (Weerasinghe and DuBois 2008). Almost all approved intense sweeteners have been discovered by accident. Tate and Lyle, however, used the Shallenberger–Acree model as a template in the firm's quest for safe, nonnutritive sweeteners with desirable taste properties. One hypothesis was to render sucrose intensely sweet by substitution of a hydroxyl group with chlorine to give it a more electronegative, hydrophobic component. The result is the widely accepted sweetener sucralose (1,6-dichloro-1,6-dideoxy-α-D-frutofuranosyl-4-chloro-4-deoxy-β-D-galactopyranoside).

Structure 5.3   Sucralose

Sugar Solubility

Sugars follow the general solubility rule of "like dissolves in like." With their multiple hydroxyl groups, they are highly soluble in water and alcohol and have low solubility in nonpolar solvents. Individual sugars show a considerable range in water solubility, from 30% to 80% by weight. Figure 5.4 shows the variability of fructose, glucose, sucrose, and lactose solubility with increasing temperature (Shallenberger and Birch 1975). Some factors that influence sugar water solubility are whether hydroxyl

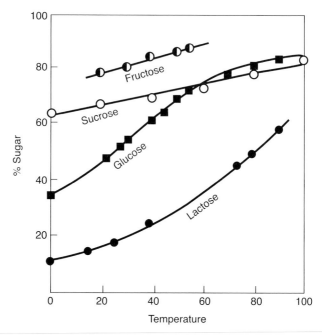

Figure 5.4   Approximate solubility of various sugars at different temperatures.
Source: **Shallenberger and Birch 1975.**

groups are oriented to be readily available to hydrogen bond with water. For example, mannose and galactose form intramolecular hydrogen
bonds and have lower water solubility than glucose. Monosaccharides
have more free hydroxyls per sugar unit than polysaccharides and are
more water-soluble. Maltose in water solution adopts the conformation so
that the more hydrophobic "undersides" of the two glucose units are oriented toward each other, giving the molecule as a whole more hydrophilic
character. (Maltose is more water-soluble than cellobiose.) Sugar monohydrate crystals tend to be less soluble than anhydrous sugars; for example,
anhydrous β-lactose is more water-soluble than α-lactose monohydrate.

Alcohol extraction is widely used for isolating sugars from plant materials. Proteins, polysaccharides, and higher oligosaccharides are insoluble in 80% ethanol and readily separated from sugars by filtration or centrifugation. As molecular weight increases, solubility of oligosaccharides
in aqueous alcohol decreases. Fractional precipitation is used industrially
to isolate maltodextrins with varying dextrose equivalencies. Mixtures of
sugars are more soluble than individual sugars; for example, a mixture of
invert sugar and sucrose has a solubility of 75.1% (w/v) at 20°C compared
with 67.7% for sucrose alone (Daniel et al. 2007). Crystalline sugars have a

Table 5.2   Heat of Solution of Crystalline Sugars and Sugar Alcohols

Compound	Heat of Solution (Kcal/g)	Source
Xylitol	−36.6	Bär 1991
Sorbitol	−26.5	Kearsley and Deis 2006
Mannitol	−28.9	Kearsley and Deis 2006
Glucose (monohydrate)	−25.3	Pancoast and Junk 1980
Glucose (anhydrous)	−14.5	Pancoast and Junk 1980
Sucrose	−3.85	Pancoast and Junk 1980

negative heat of solution, losing heat and providing a cooling effect when dissolved in water. Table 5.2 lists the heats of solution for selected crystalline sugars and sugar alcohols. The pleasant sensation when xylitol and other sugar alcohols dissolve in the mouth finds application in chewing gums and mint candies.

Crystallinity of Sugars

The **crystalline** state can be defined as **an ordered arrangement of molecules**. It involves a three-dimensional ordered array of molecules in which there is periodicity and symmetry (Flink 1983); all molecules are equivalent with respect to their binding energy. Individual sugars can exist in two or more allomorphic forms, similar to carbon having the allotropic forms of diamond, graphite, and fullerenes. The opposite of crystalline is the **amorphous** state, the absence of a regular structure. Crystalline materials have a sharp melting point, and in x-ray diffraction a uniform pattern of scattering will be presented. The tendency of sugars to crystallize is inversely related to solubility. Highly soluble sugars are inclined to be difficult to crystallize, and the opposite is characteristic of sugars having low solubility. Reducing sugars are more difficult to crystallize than nonreducing sugars because of the presence of different anomeric and ring forms. Sugar refining is essentially a process of purification by crystallization. Highly concentrated sucrose solutions are brought about by water evaporation. Presence of other sugars and impurities will interfere and prevent crystallization. Sucrose inversion is prevented in sugar refining by adjusting the pH to alkaline conditions. Corn syrup can prevent the crystallization of glucose in jams and jellies, and corn syrup is frequently added to frozen desserts to prevent sucrose crystallization. In hard candy manufacture, corn syrup will be used in conjunction with sucrose to form an amorphous glass and prevent crystallization.

Table 5.3   Percentage of Water Absorbed by Sugars From Moist Air[a]

Compound	60% RH[b] (9 days)	100% RH (25 days)
D-Glucose, crystalline	0.07	14.5
D-Fructose, crystalline	0.63	73.4
Sucrose	0.03	18.4
Invert Sugar	3.0	74.0
β-Maltose hydrate	5.1	
β-Lactose, anhydrous	1.2	1.4
α-Lactose hydrate	5.1	

[a]Modified from Hodge and Osman 1976.
[b]RH = relative humidity at 20°C.

Hygroscopicity

Hygroscopicity is the rate and extent of water absorption from the atmosphere. It is easily measured by exposing a substance to a given relative humidity (RH) and then measuring the increase in weight. Table 5.3 gives hygroscopicity measurements for some selected sugars. Amorphous sugars are "water-seeking" and more hygroscopic than crystalline sugars. Water uptake increases rapidly above 58% RH for D-fructose, 80% RH for D-glucose, and 83% RH for sucrose (Hodge and Osman 1976). Figure 5.5 shows water absorption versus water activity for amorphous maltose (curve 1) and crystalline maltose (curve 2). Anhydrous crystalline

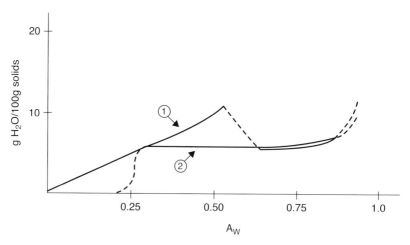

Figure 5.5   Water vapor sorption by amorphous (curve 1) and crystalline (curve 2) maltose.
Source: Flink 1983.

maltose takes up one molecule of water at a water activity (A_w) of 0.25, where it changes to crystalline maltose monohydrate. Amorphous maltose crystallizes as crystalline maltose monohydrate at an approximate A_w of 0.6.

Humectancy

A **humectant** is a substance that has the ability to resist changes in moisture content. Humectants are food ingredients that will attract water and maintain plasticity in a food system, without producing "hygroscopic" effects. High-molecular-weight carbohydrates and noncrystallizing sugars are good humectants. Amorphous sucrose in brown sugar is ineffective as a humectant because it will crystallize, lose its affinity for water, and result in loss of plasticity. In product development, selection of an appropriate humectant for a food system is usually done on an empirical basis.

Viscosity

Viscosity is defined as resistance to flow. There are a number of viscometers that give fundamental measurements of viscosity (e.g., the Brookfield and Ostwald viscometers). Viscosity is dependent on concentration and molecular size and shape. Increasing temperature will result in reduced viscosity. A 10% sugar concentration provides "body" to soft drinks, and sugar solutions at 40% and above are highly viscous. Sucrose solutions will have a measurable higher viscosity than invert sugar solutions, if both are at equivalent concentrations, because of sucrose's larger molecular size. Honey is highly viscous because of its high sugar concentration, >80% sugar by weight, c.a. 17% water content. Low DE corn syrups and maltodextrins increase viscosity and are bodying agents for food systems. High-molecular-weight polysaccharides are used to provide body to low-calorie soft drinks.

Freezing Point Depression and Boiling Point Elevation

A 1M sugar solution will lower the freezing point by 1.86°C and raise the boiling point by 0.56°C. Thus, sugars will lower the freezing point of frozen desserts by 2–5°C. Replacement of sucrose with corn syrup will result in a lower freezing point and softer texture. In candy making, increase in sugar concentration from water loss results in an increase in the boiling point. Measurement of boiling temperature becomes a measurement of sugar concentration required for different confections. For example, the final cooking temperature recommended for peanut brittle is 300°F (149°C). Measurement of the freezing point of milk has wide acceptance in the dairy industry for detection of water addition (Wehr and Frank 2004).

Osmotic Effects

When dissolved in water, sugars will decrease vapor pressure and water activity (A_w). The impact on A_w can be estimated from the following equation:

$$A_w = \frac{\text{moles water}}{\text{moles water} + \text{moles solute}} \tag{5.4}$$

Microbial growth will be inhibited when A_w is lowered sufficiently. As a generality, bacteria will be inhibited at an A_w of 0.81 or below, yeasts at 0.76 or below, and molds at 0.70 or below. Jams, jellies, and preserves are examples of food preservation by reduction of A_w through sugar addition, one of the oldest methods of preserving foods.

Vocabulary

Crystalline—An ordered arrangement of molecules
Amorphous—Noncrystalline; devoid of regular structure
Hygroscopicity—The rate and extent of water absorption from the atmosphere
Humectant—A substance that has the ability to resist changes in moisture content
Viscosity—Resistance to flow

References

Acree TE, Lindley M. 2008. Structure-activity relationship and AH-B after 40 years. In: Weerasinghe DK, DuBois GE, editors. *Sweetness and sweeteners: biology, chemistry and psychophysics*. ACS Symposium Series 979. Washington D.C.: American Chemical Society, pp. 96–108.

Bär A. 1991. Xylitol. In: Nabors LO, Gelardi RC, editors. *Alternative sweeteners, 2nd ed*. New York: Marcel Dekker, pp. 349–379.

Birch GG, Lee CK. 1974. Structural functions of taste in the sugar series: sensory properties of deoxy sugars. *J Food Sci* 39:947–949.

Daniel JR, Yao Y, Weaver C. 2007. Carbohydrates: Functional properties. In: Hui YH, editor. *Food chemistry: principles and applications, 2nd ed.* Sacramento, CA: Science Technology System, pp. 5(1)–5(26).

Flink JM. 1983. Structure and structure transitions in dried carbohydrate materials. In: Peleg M, Bagley EB, editors. *Physical properties of foods.* Westport, CT: AVI Publishing, pp. 473–521.

Hodge JE, Osman EM. 1976. Carbohydrates. In: Fennema OR, editor. *Principles of food science. Part I: Food chemistry.* New York: Marcel Dekker, pp. 41–138.

Kearsley MW, Deis RC. 2006. Sorbitol and mannitol. In: Mitchell H, editor. *Sweeteners and sugar alternatives in food technology.* Ames, IA: Blackwell Publishing, pp. 249–261.

Lindley M. 2006. Other sweeteners. In: Mitchell H, editor. *Sweeteners and sugar alternatives in food technology.* Ames, IA: Blackwell Publishing, pp. 331–360.

Nelson AL. 2000. *Sweeteners: alternative.* St. Paul, MN: Eagan Press.

Pancoast HM, Junk WR. 1980. *Handbook of Sugars, 2nd ed.* Westport, CT: AVI Publishing.

Shallenberger RS. 1963. Hydrogen bonding and the varying sweetness of the sugars. *J Food Sci* 28:584–589.

Shallenberger RS. 1993. *Taste chemistry.* New York: Blackie Academic & Professional.

Shallenberger RS. 1998. Sweetness as a sensory property. In: Alexander RJ, editor. *Sweeteners: nutritive.* St. Paul, MN: Eagan Press, pp. 9–16.

Shallenberger RS, Acree TE. 1967. Molecular theory of sweet taste. *Nature* 216:480–482.

Shallenberger RS, Acree TE. 1969. Molecular structure and sweet taste. *J Agric Food Chem* 17:701–703.

Shallenberger RS, Birch GG. 1975. *Sugar chemistry.* Westport, CT: AVI Publishing.

Skytte UP. 2006. Tagatose. In: Mitchell H, editor. *Sweeteners and sugar alternatives in food technology.* Ames, IA: Blackwell Publishing, pp. 262–294.

Weerasinghe DK, DuBois GE. 2008. *Sweetness and sweeteners: biology, chemistry and psychophysics.* ACS Symposium Series 979. Washington D.C.: American Chemical Society.

Wehr HM, Frank JF. 2004. *Standard methods for the examination of dairy products, 17th ed.* Washington, D.C.: American Public Health Association.

6 Analytical Methods

Introduction

This chapter is not a comprehensive review of food carbohydrate analytical methods. The intent is to describe the underlying principles of the most widely used methods for analysis of sugars and to briefly describe their sensitivity and limitations. This should enable one to make appropriate choices in methodology for specific applications. The selected topics and case studies reflect the author's experience and bias in the analysis of fruit juices and other beverages.

Detailed description of analytical methods will not be given. Rather, referral will be made to several excellent references. An excellent overview of carbohydrate methodology is the chapter by James BeMiller in the popular text *Food Analysis* (Nielsen 2010). AOAC's *Official Methods of*

Food Carbohydrate Chemistry, First Edition. Ronald E. Wrolstad.
© 2012 John Wiley & Sons, Inc. Published 2012 by John Wiley & Sons, Inc.

Analysis (Horwitz and Latimer 2010) is widely regarded as the standard source for analytical methods. All of the approved AOAC methods undergo a rigorous collaborative study to ensure their validity and repeatability by different analysts. Legal departments will often prefer that official standard analytical methods be used, particularly if litigation is a possible issue. *Determination of Food Carbohydrates* is an excellent monograph by Southgate (1991) that gives a lucid and practical description of food carbohydrate methods. It includes a chapter that has concise descriptions of procedures for selected methods. *Methods in Carbohydrate Chemistry* (Whistler and Wolfrom 1962, 1963; Whistler 1963, 1964, 1965; Whistler and BeMiller 1972, 1976, 1980; BeMiller and Whistler 1993; BeMiller 1994) are a series of monographs that contain chapters written by experts in the field on common and specialized methods. *Carbohydrates,* Section E in *Handbook of Food Analytical Chemistry* (Wrolstad et al. 2005) gives detailed instructions along with tips regarding time considerations, stages in a procedure where a sample can be stored, critical parameters, etc. Professional societies and trade associations representing various food commodities have a vested interest in standard analytical methods. The American Association of Cereal Chemists (AACC) first published *Methods for the Analysis of Cereals and Cereal Products* in 1922. The 11th Edition, *AACC Intl. Approved Methods of Analysis,* is available online (AACC International 2010). Two popular books for wine analysis are *Principles and Practices of Winemaking* (Boulton et al. 1996) and *Wine Analysis and Production* (Zoecklein and others 1995). The American Society of Brewing Chemists (ASBC) publishes a manual on *Methods of Analysis* with frequent updates that is now available on CD-ROM (ASBC 2009). *Standard Methods for the Examination of Dairy Products* (Wehr and Frank 2004) is widely used by the dairy industry.

Physical Methods

Refractometry

The refractive index of aqueous sugar solutions varies linearly with concentration (Figure 6.1); thus, sugar concentration can conveniently be measured as a function of refractive index. Bench-top and hand-held refractometers have widespread usage in industry, with most instruments being calibrated for percent soluble solids or °Brix. °Brix was previously defined (Chapter 2) as the percent sugar by weight in an aqueous solution.

$$°\text{Brix} = \frac{\text{wt sugar}}{\text{wt sugar} + \text{wt water}} \times 100 \tag{6.1}$$

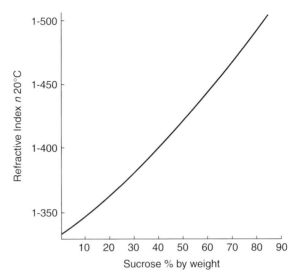

Figure 6.1   Relationship between refractive index and sucrose concentration.
Source: **Southgate 1991.**

Because the refractive index of sucrose solutions is used for calibration of refractometers, accurate measurements are limited to sucrose solutions. In practice, the measurement is widely used for fruit juices and other beverages in which other dissolved substances, such as organic acids, will also contribute to the refractive index. Temperature needs to be controlled for accurate measurements. Portable hand-held refractometers are used in the vineyard to monitor when grapes are ready for harvest. Alcohol has a marked effect on refractive index; hence, refractometers cannot be used for measuring sugar levels in wine and beer. (Alcohol's impact on refractive index also contributes to the attractive appearance of a flute of champagne, a glass of Chardonnay, or a snifter of brandy.)

The Pearson square is a useful arithmetic diagram for determining the proportions to be used when mixing or diluting sugar solutions to achieve a desired concentration. Refer to Figure 6.2 where the objective is to dilute a 75 °Brix corn syrup solution with water to 15 °Brix. To solve this problem, the desired °Brix (15°) is placed in the center of the square and the °Brix of the most concentrated sample (75 °Brix) is in the upper left-hand corner. The °Brix of the other solution (water = 0°) is placed in the lower left-hand corner. The desired value (15°) is subtracted diagonally from the starting materials. Fifteen parts (by weight) of the 75 °Brix corn syrup diluted with 60 parts water (by weight) will give a 15 °Brix solution.

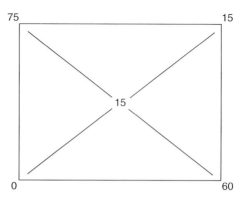

Figure 6.2   Diagram illustrating the Pearson square method for determining weight proportions of ingredients to obtain desired °Brix.

Density

A linear relationship holds for specific gravity and concentration of aqueous sucrose solutions (Figure 6.3). Determination of specific gravity with a pycnometer is one means of determining sugar concentration. This process is somewhat tedious. The use of hydrometers, however, that are

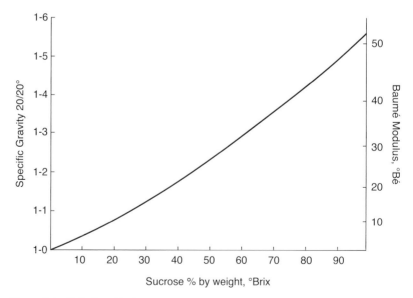

Figure 6.3   Relationship between specific gravity and sucrose concentration. *Source:* **Southgate 1991.**

calibrated in specific gravity, °Brix, or °Baumé is rapid and convenient. There are °Baumé scales for liquids both heavier and lighter than water. Hydrometers find usage in brewing, wine making, dairy processing, sugar refining, and the starch industry.

Polarimetry

Optical rotation of sugars and the use of polarimeters for identifying sugars, examining mutarotation phenomena, and measuring sugar concentration are discussed in Chapter 1. The Appendix includes a laboratory exercise where determination of specific rotation is used for identification of unknown sugar solutions and for demonstrating mutarotation and optical activity. Use of polarimeters for measuring sugar concentration is limited to pure sugar solutions and not applicable to mixtures of different sugars. Polarimeters (also known as saccharimeters) have been used in sucrose refineries for measuring sugar concentration since the mid-nineteenth century.

Colorimetric Methods

Sugars are colorless but reactive compounds. As their separation by paper and thin-layer chromatography evolved, a number of creative techniques were developed for their visual detection. Treatment with acids of varying strength could lead to hydrolysis of glycosidic linkages and sugar dehydration (refer to Chapter 3). Subsequent complexation with various compounds can generate colored substances, permitting visualization and quantitation. Qualitative detection of hexoses versus pentoses and aldoses versus ketoses is possible. Many of these same reactions are used for spectrophotometric determination of sugar concentration. A brief description of some of the more common spectrophotometric procedures will be given, along with referral to other sources for detailed instructions.

Total Sugars by Phenol-Sulfuric Acid

In this procedure, concentrated sulfuric acid is added to an aqueous sugar sample that also contains phenol. Sugar dehydration products complex with phenol to generate a yellow color that can be measured at 490 nm.

Some useful references include the following: Southgate 1991; Fournier 2005; BeMiller 2010. Accuracy is $\pm 2\%$, and sensitivity is in the order of 5 μg (Fournier 2005).

Reducing Sugar Methods

The oxidation of reducing sugars by metal ions is discussed in Chapter 3, and an experiment using the Fehlings test for qualitative determination of reducing sugars is included in the Appendix. The **Lane–Eynon Method** is an extension of the Fehlings reaction to determine the quantity of reducing sugars by titration. It is commonly used in wineries to measure total reducing sugars. A standard glucose solution is reacted with alkaline copper sulfate under specified heating conditions. One milliliter of wine is added to a second standard glucose solution, and the amount of sugar required for reduction is determined titrametrically using methylene blue as an endpoint indicator (Zoecklein et al. 1995; Boulton et al. 1996). The difference between the two solutions is the reducing sugar content of the wine.

In the **Somogi–Nelson Method** for total reducing sugars, the sample is reacted with alkaline copper sulfate, followed by addition of an arsenomolybdate complex. The cuprous ions formed reduce the arsenomolybdate complex to form a stable blue color that is measured spectrophotometrically (Southgate 1991; Fournier 2005; BeMiller 2010). Accuracy and sensitivity is similar to the phenol-sulfuric acid procedure (Fournier 2005).

Chromatographic Methods

Food extracts will typically contain a mixture of sugars, and the analyst is usually interested in identifying the individual sugars and measuring their concentration. Chromatographic methods offer the advantages of removing interfering substances and providing retention indices that are helpful for identification. Chromatographic methods are rapid, sensitive, and suitable for routine analyses.

Paper and Thin-Layer Chromatography

Paper and thin-layer chromatography were widely used for analysis of sugars in the 1950s and 1960s. Equipment costs are minimal, and sensitivity is in the order of 1 μg. A number of visualization sprays have been developed, some of which can distinguish reducing sugars from nonreducing sugars, pentoses from hexoses, and aldoses from ketoses (Stahl 1969; Southgate 1991). Although its usage has been replaced in most laboratories with high-performance liquid chromatography (HPLC), it is still a very useful technique. For example, Swallow and Low (1993) judiciously used paper chromatography in conjunction with HPLC, gas–liquid chromatography (GLC), and ^{13}CNMR in identifying trace oligosaccharides in commercial beet medium invert syrup. Dinitrosalicylic acid reagent was used to determine whether the separated oligosaccharides were reducing or nonreducing, and anthrone reagent was used to determine whether they were ketose-containing. Stahl (1969) is a comprehensive handbook on thin-layer chromatography.

Gas–Liquid Chromatography

Sugars are nonvolatile and need to be derivatized before they can be analyzed by GLC. Formation of alditol acetates is one procedure that is widely used in the separation, identification, and quantitation of the sugars in cell wall polysaccharides (Melton and Smith 2005). Following hydrolysis of the polysaccharides, the sugars are reduced to alditols with sodium-borohydride, and then all hydroxyl groups are acetylated using acetic anhydride. Figure 6.4 illustrates the reduction of D-fructose to D-mannitol and D-glucitol, followed by their acetylation to produce the alditol acetate derivatives. Figure 6.5 is a chromatogram of 13 sugar standards. A limitation of the method is that glucose, fructose, and sorbitol will all form D-glucitol hexaacetate, and ketose sugars such as fructose will form two alditols.

Trimethylsily sugar derivatives are formed by reacting sugars with hexamethyldislazane (HMDS) in presence of hexane and pyradine under anhydrous conditions. The reaction will not go to completion if any water is present, so sugars and sugar extracts need to be in a dried form. (Refer to the reaction of β-D-glucopyranose shown in Chapter 3.) Crystalline β-D-glucopyranose will form one peak: trimethylsilyl-2,3,4,6-O-tetratrimethylsilyl-β-D-glucopyranoside. (Refer to Chapter 1 regarding nomenclature.) Aqueous sugar samples, when dried, will be present in different anomeric and ring forms, each producing a different trimethylsilyl derivative. This feature permits quantitative examination of complex mutarotation. Trimethylsily derivitization of sugar mixtures will give very complex chromatograms; hence, alditol acetate derivatives are preferable for analysis of cell wall polysaccharides. The method is very sensitive with detection in the order of 20 ng.

98

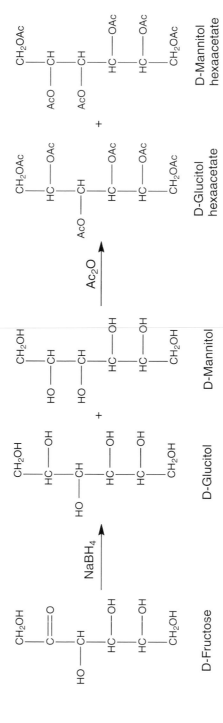

Figure 6.4 Scheme showing reduction of D-fructose to D-glucitol (sorbitol) and D-mannitol followed by acetylation to form the alditol acetate derivatives.

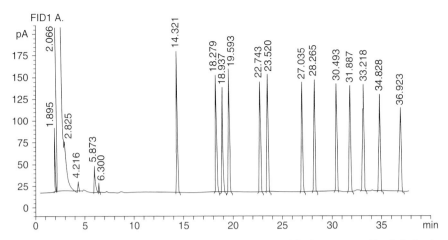

Figure 6.5   Chromatogram of a standard mixture of alditol acetates. Erythritol triac-etate (peak 14.321), 2-deoxyribitol tetraacetate (peak 18.279), rhamnitol pentaacetate (peak 18.937), fucitol pentaacetate (peak 19.593), ribitol pentaacetate (peak 22.743), arabinitol pentaacetate (peak 23.520), xylitol pentaacetate (peak 27.075), 2-deoxyglucitol hexaacetate (peak 28.265), allitol hexaacetate (peak 30.493), mannitol hexaacetate (peak 31.887), galacti-tol hexaacetate (peak 33.218), glucitol hexaacetate (peak 34.828), *myo*-inositol hexaacetate (peak 36.923).
Source: used with permission from Melton and Smith 2005.

Case Study

Use of Trace Oligosaccharide Analysis in Determining Fruit Juice Authen-ticity: Nicholas Low (1998) took advantage of the sensitivity of GLC of sugar trimethylsily derivatives to detect undeclared addition of inexpen-sive commercial sweeteners in fruit juices. High-fructose corn syrup con-tains trace amounts of isomaltose and maltose. Medium and total invert syrup is produced by acid or enzymic hydrolysis of cane or beet sugar. A carbonium ion is produced during sucrose hydrolysis, which will read-ily combine with other sugar molecules. Four fingerprint oligosaccharides were characterized as trisaccharides formed from addition of the fructo-furanosyl oxonium ion to sucrose. These sugars are known as reversion sugars (refer to Chapter 3). Detection of these compounds by GLC of trimethylsily derivatives was quickly adopted for detecting adulteration of fruit juices with high-fructose corn syrup or invert beet/cane syrup. Low and Hammond (1996) detected two unidentified peaks in commercial apple juice samples that they characterized as fructose disaccharides originating from hydrolyzed inulin syrup. Thus, there are trace oligosaccharides that serve as marker compounds for the presence of high-fructose corn syrup, invert beet/cane syrup, and high-fructose syrup from inulin.

HPLC

In most laboratories, HPLC has become the method of choice for sugar analysis. It offers the advantages of being quantitative and rapid, with typical runs being 20 minutes or less. For many applications, little sample preparation is required (e.g., filtration of a fruit juice through a Millipore filter before injection). Refractive index detectors with sensitivity in the order of 1 mg are most commonly used since UV detection is impractical because of possible interference in the low UV range, where sugars absorb. For many food samples, sugar concentration is such that sensitivity is not an issue. A limitation of refractive index detectors is that gradient elution cannot be used. Normal-phase chromatography with amino-bonded columns is widely used for sugar analysis. Figure 6.6 shows a

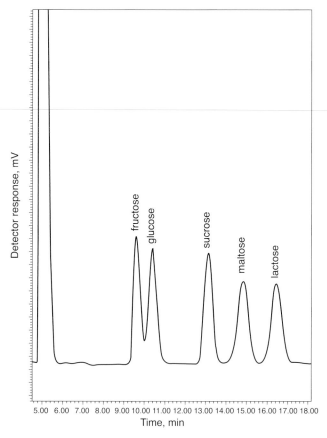

Figure 6.6   Chromatogram of sugar standards. Column: 250 × 4.6 mm amino-propylsilyl column (Supelco); mobile phase, 7:3 acetonitrile/water, isocratic; flow rate, 1.0 mL/min; ambient column temperature.
Source: used with permission from Ellefson 2005.

chromatogram of sugar standards (Ellefson 2005). Elution order is monosaccharides and sugar alcohols, followed by disaccharides and higher oligosaccharides. A problem with amino-bonded columns is that reducing sugars can condense with the amino function. (This is the initial reaction in the Maillard browning scheme discussed in Chapter 4.) This shortens column life and loss of some of the sugar being measured. Another problem is poor resolution of glucose, galactose, and sorbitol. Accurate measurement of glucose and sorbitol is particularly important for fruit juices.

Calcium-loaded cation exchange resin columns give good resolution of monosaccharides and sugar alcohols. Figure 6.7 shows the separation of

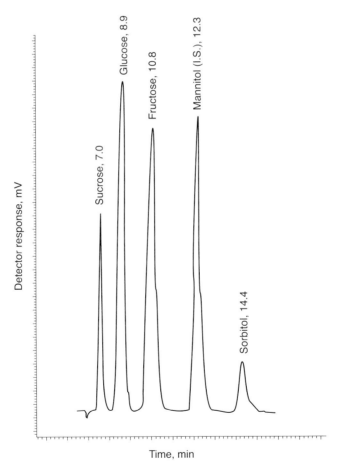

Time, min

Figure 6.7   HPLC chromatogram of sugar standards. Column: 300 × 7.8 mm Bio-Rad Aminex HPX-87C column; mobile phase, 200 Ca(No$_3$)$_2$ in distilled water; column temperature, 85°C; flow rate, 0.7 mL/min.
Source: **used with permission from Wrolstad 1993.**

sugar standards using water at 85°C as the mobile phases. The mechanism of separation is based on the strength of bonding between *cis*-glycols of sugars with the Ca^{+2} loaded on the column. Oligosaccharides are eluted first because the larger mass will more readily be swept away by the mobile phase. Elution order is related to the number and strength of *cis*-glycol complexing. Acyclic polyhydric alcohols such as sorbitol, xylitol, and mannitol are late eluters. Presumably this is because the free rotation of the acyclic molecule allows for optimum *cis*-glycol binding to Ca^{+2}. Sugars can also be separated on anion-exchange (AE) columns under alkaline conditions. Sugars are weak acids and will be anions at high (12–14) pH. AE-HPLC columns are typically used in conjunction with sensitive electrochemical detectors for analysis of complex mixtures of food sugars.

Enzymic Methods

Enzyme kits are commercially available that permit the spectrophotometric determination of individual sugars spectrophotometrically (Henniger 1998). In the determination of glucose, glucose in the presence of ATP is phosphorylated to glucose-6-phosphate. Glucose-6-phosphate is subsequently converted to 6-phosphogluconate by glucose-6-phosphate dehyrogenase in the presence of NADP.

$$glucose + ATP \xrightarrow[hexokinase]{} glucose\text{-}6\text{-}phosphate + ADP \qquad (6.2)$$

$$glucose\text{-}6\text{-}phosphate + NADP \xrightarrow[glucose\text{-}6\text{-}phosphatedehyrogenase]{}$$
$$6\text{-}phosphogluconate + NADPH + H^+ \quad (6.3)$$

The formation of NADH is measured at 334 nm, and the quantity is stoichiometric with glucose concentration. The enzyme kits contain the enzymes, cofactors, and buffer salts required for the reaction. Kits are available for the determination of fructose, lactose, lactulose, maltose, sucrose, sorbitol, and trehalose (Megazyme 2010). Enzyme kits are effective for determining starch content of food products.

Anecdote

Enzymic Analysis of L-Malic Acid as Critical Evidence in the Beech-Nut Adulterated Apple Juice Scandal (Wrolstad and Durst 2006). Refer to this landmark 1988 case discussed in Chapter 2. Key analytical information was determination of L-malic acid by enzymic analysis contrasted with determination of malic acid by HPLC. Plants synthesize L-malic acid and not the D-isomer. HPLC gave values twice that of enzymic analysis, since the D-isomer in synthetic DL-malic would not be measured by enzymic assay.

Carbon Stable-Isotopic Ratio Analysis (SIRA)

Through photosynthesis, plants convert atmospheric CO_2 to sugars. However, all plants do not use the same pathway for assimilation of CO_2. The Calvin cycle is the most common pathway, where ribulose-5-phosphate combines with CO_2 to form 3-phosphoglyceric acid. Plants following the Calvin cycle are also known as C-3 plants. In the Hatch–Slack pathway, CO_2 fixation occurs when phosphoenol pyruvate combines with CO_2 to form oxaloacetate. Plants following this C-4 pathway tend to be of tropical origin. There are two stable carbon isotopes, the natural abundance of carbon-12 being 99% and carbon-13 being 1%. All plants discriminate in favor of the lighter isotope, but Calvin cycle plants discriminate more. The proportions of ^{13}C and ^{12}C in plant materials can be measured by high-resolution mass spectrometry. The material is combusted and the proportions of ^{13}C and ^{12}C in the CO_2 produced is measured and compared with that of a standard, PeeDee Belemnite (PDB), a fossil calcium carbonate from South Carolina. Carbon stable isotope ratios are reported as values of δ (DEL), which are differences in parts per thousand (per mil) between the isotopic composition of the sample and the PDB standard.

$$\delta^{13}C(\%) = \frac{(^{13}C/^{12}C)_{sample} - (^{13}C/^{12}C)_{PDB}}{(^{13}C/^{12}C)_{PDB}} \times 1000 \tag{6.4}$$

Sugar beets use the Calvin cycle for CO_2 fixation, whereas cane sugar uses the Hatch–Slack cycle. Thus, carbon SIRA is a means for determining whether sucrose is derived from beet or cane (Bricout and Fontes 1974). Table 6.1 lists the $\delta^{13}C/^{12}C$ values for selected sweeteners and fruit juices.

Table 6.1 $\delta^{13}C/^{12}C$ Values of Sweeteners and Fruit Juices[a]

Sample	$\delta^{13}C/^{12}C$ (‰)	Range
Cane sugar	−11.4	−11.2 to −11.7
Beet sugar	−25	
Corn syrup	−9.7	−9.5 to −9.8
Honey[b] ($n = 84$)	−25.2	−23.4 to −26.8
Apple juice ($n = 40$)	−25.3	−22.5 to −9.8
Orange juice ($n = 38$)	−24.5	−23.4 to −25.6
Grape juice	−26.7	−23.5 to −30.5
Pineapple juice	−12.2	−11.2 to −13.5

Sources: [a]Krueger 1998; [b]Doner and White 1977.

> **Case Study**
>
> Case study—Application of SIRA in Determining Authenticity. High-fructose corn syrup is inexpensive, has a sugar profile similar to honey and many fruit juices, and has no distinctive or disagreeable taste. This made it an attractive adulterant for several food products in which unscrupulous individuals and companies wanted to turn a quick profit. Honey was one of the first products to be analyzed by carbon SIRA (Doner and White 1977). Almost all floral honeys are derived from Calvin cycle plants and have $\delta^{13}C/^{12}C$ values near $-25‰$ (Table 6.1). Doner et al. (1979) effectively used carbon SIRA to determine that imported candied pineapple and papaya fruit were processed with cane or corn syrup rather than honey. In 1978, Krueger Enterprises analyzed a large sampling of commercial apple juices from the New England area and found widespread undeclared addition of corn or cane syrup (Krueger 1984). The Processed Apple Institute took prompt action and implemented an industry-wide carbon SIRA testing. Carbon SIRA was widely adopted as a parameter for detecting undeclared addition of cane or corn syrup to fruit juice (Krueger 1988).

Note: Included in this chapter are brief presentations of case studies where "new" and more sensitive methods were used to detect commercial fraud. Crafty firms that cheat will operate somewhat below the level of sensitivity of methods that are in place for detection of undeclared ingredients. Regulatory agencies and trade associations have the continuing challenge of developing more sensitive and more rapid methods of detection.

References

AACC International. 2010. *Approved methods of analysis, 11th ed.* St. Paul, MN: American Association of Cereal Chemists.

ASBC. 2009. Methods of analysis CD-ROM. St. Paul, MN: American Society of Brewing Chemists.

BeMiller JN, Whistler RL. 1993. *Methods in carbohydrate chemistry. Vol IX: Lipopolysaccharides, separation and analysis, glycosylated polymers.* New York: John Wiley & Sons.

BeMiller JN. 1994. *Methods in carbohydrate chemistry. Vol X: Enzymic methods.* New York: John Wiley & Sons.

BeMiller JN. 2010. Carbohydrate analysis. In: Nielsen SS, editor. *Food analysis, 4th ed.* New York: Springer, pp. 147–177.

Boulton RB, Singleton VL, Bisson LF, Kunkee RE. 1996. *Principles and practices of winemaking.* New York: Chapman and Hall.

Bricout J, Fontes JC. 1974. Analytical distinction between cane and beet sugar. *Ann Fals Exp Chim* 67:211–215.

Doner LW, Chia D, White JW Jr. 1979. Mass spectrometric ^{13}C/^{12}C determinations to distinguish honey and C-3 plant sirups from C-4 plant sirups (sugar cane and corn) in candied pineapple and papaya. *J Assoc Off Anal Chem* 62:928–930.

Doner LW, White JW Jr. 1977. Carbon-13/Carbon-12 ratio is relatively uniform among honeys. *Science* 1997:891–892.

Ellefson, W. 2005. HPLC of mono-and disaccharides using refractive index detection. Unit E1.2.1. In: Wrolstad RE, Reid DS, Smith DM, Penner MH, Decker EA, Sporns P, editors. *Handbook of food analytical chemistry.* Hoboken, NJ: John Wiley & Sons, pp. 661–669.

Fournier E. 2005. Colorimetric quantification of carbohydrates. Unit E1.1.1. In: Wrolstad RE, Reid DS, Smith DM, Penner MH, Decker EA, Sporns P, editors. *Handbook of food analytical chemistry.* Hoboken, NJ: John Wiley & Sons, pp. 653–660.

Henniger G. 1998. Enzymic methods of food analysis. In: Ashurst PR, Dennis MJ, editors. *Analytical methods of food authentication.* New York: Blackie Academic & Professional, pp. 137–181.

Horwitz W, Latimer GW Jr. 2010. *Official methods of analysis of AOAC International. 18th ed., Revision 3.* Gaitherburg, MD: AOAC International.

Krueger DA. 1988. Applications of stable isotope ratio analysis to problems of fruit juice adulteration. In: Nagy S, Attaway JA, Rhodes ME, editors. *Adulteration of fruit juice beverages.* New York: Marcel Dekker, pp. 109–124.

Krueger DA. 1998. Stable isotope analysis by mass spectrometry. In: Ashurst PR, Dennis MJ, editors. *Analytical methods of food authentication.* New York: Blackie Academic & Professional, pp. 14–35.

Krueger HW. 1984. Detection of adulteration and fraud in food products using MS. *Am Lab* 16:90–91.

Low NH. 1998. Oligosaccharide analysis. In: Ashurst PR, Dennis MJ, editors. *Analytical methods of food authentication.* New York: Blackie Academic & Professional, pp. 97–136.

Low NH, Hammond DA. 1996. Detection of high fructose syrup from inulin in apple juice by capillary gas chromatography with flame ionization detection. *Fruit Process* 4:135–139.

Megazyme. 2010. *Advanced bio-analysis test kits for the food, feed, fermentation, wine, brewing and dairy industries.* Available from: http://secure.megazyme .com/homepage.aspx. Accessed October 7, 2010.

Melton LD, Smith BG. 2005. Determination of neutral sugars by gas chromatography of their alditol acetates. Unit E3.2 In: Wrolstad RE, Reid DS, Smith DM, Penner MH, Decker EA, Sporns P, editors. *Handbook of food analytical chemistry.* Hoboken, NJ: 0 John Wiley & Sons, pp. 721–733.

Nielsen SS. 2010. *Food analysis, 4th ed.* New York: Springer.

Southgate DAT. 1991. *Determination of food carbohydrates, 2nd ed.* New York: Elsevier Applied Science.

Stahl G. 1969. *Thin-layer chromatography: a laboratory handbook, 2nd ed.* [English translation]. New York: Springer-Verlag.

Swallow KW, Low NH. 1993. Isolation and identification of oligosaccharides in a commercial beet medium invert syrup. *J Agric Food Chem* 41:1587–1592.

Wehr HM, Frank JF. 2004. *Standard methods for the examination of dairy products, 17th ed.* Washington, D.C.: American Public Health Association.

Whistler RL. 1963. *Methods in carbohydrate chemistry. Vol 3: Cellulose.* New York: Academic Press.

Whistler RL. 1964. *Methods in carbohydrate chemistry. Vol 4: Starch.* New York: Academic Press.

Whistler RL. 1965. *Methods in carbohydrate chemistry. Vol 5: General polysaccharides.* New York: Academic Press.

Whistler RL, BeMiller JN. 1972. *Methods in carbohydrate chemistry. Vol 6: General carbohydrate methods.* New York: Academic Press.

Whistler RL, BeMiller JN. 1976. *Methods in carbohydrate chemistry. Vol 7: General methods, glycosaminoglycans, and glycoproteins.* New York: Academic Press.

Whistler RL, BeMiller JN. 1980. *Methods in carbohydrate chemistry. Vol 8: General methods.* New York: Academic Press.

Whistler RL, Wolfrom ML. 1962. *Methods in carbohydrate chemistry. Vol 1: Analysis and preparation of sugars.* New York: Academic Press.

Whistler RL, Wolfrom ML. 1963. *Methods in carbohydrate chemistry. Vol 2: Reactions of carbohydrates.* New York: Academic Press.

Wrolstad RE. 1993. *Analysis of sugars in fruit products and juices.* In: Proceedings of SPRI workshop on analysis of sugars in foods. September 30, 1992. New Orleans, LA: Sugar Processing Research Institute.

Wrolstad RE, Durst RW. 2006. Fruit juice authentication: What have we learned? In: Ebeler SE, Takeoka GR, Winterhalter P, editor. *Authentication of food and wine.* ACS Symposium Series 952. Washington, D.C.: American Chemical Society, pp. 147–163.

Wrolstad RE, Reid DS, Smith DM, Penner MH, Decker EA, Sporns P, editors. 2005. *Handbook of food analytical chemistry.* Hoboken, NJ: John Wiley & Sons.

Zoecklein BW, Fugelsang KC, Gump BH, Nury FS. 1995. *Wine analysis and production.* New York: Chapman and Hall.

7 Starch in Foods

Andrew S. Ross
Department of Crop and Soil Science / Department of Food Science and Technology, Oregon State University, Corvallis, Oregon

Food Carbohydrate Chemistry, First Edition. Ronald E. Wrolstad.
© 2012 John Wiley & Sons, Inc. Published 2012 by John Wiley & Sons, Inc.

Introduction

Starch is everywhere. The most abundant source of calories for humans, starch profoundly influences the texture of scores of foods, including such staples as breads, noodles, pasta, and rice. Despite being made up only of D-glucose, starch is not uniform and occurs in many natural and induced variants. Variations include differences occurring between species, -between genotypes within species, and starches with chemical or physical modifications. Given its ubiquity, it is essential that food technologists have a proficient working knowledge of starch, including the following attributes:

- How and where it occurs in nature
- Its chemical composition
- Starch granule size and structure
- How starch interacts with co-solutes and hydrolytic enzymes
- How it can be modified
- And, finally, how these factors combine to modulate the response of starch to thermal processing, especially in the presence of water

Sources of Starch

All the major economic sources of starch are plants. Starch and starch-like molecules can also be found in other kingdoms of life, including bacteria, algae, and animals (Shannon 2009). In animals, glycogen is a smaller structural analogue of the branched component of starch, amylopectin. Anywhere it is found, the physiological role of starch is to store energy. In plants, starch is mobilized by a series of hydrolytic enzymes. In germination, the initial hydrolysis products (maltose and malto-oligosaccharides) are eventually hydrolyzed to glucose to provide energy for the young growing plants prior to the onset of photosynthesis. Glucose is further converted to a hexose phosphate for use in cellular metabolism or conversion to sucrose for energy transport (Smith et al. 2005; Zeeman et al. 2010). Starch accumulates at high concentrations in reproductive structures like cereal grains (e.g., wheat, rice, maize, barley, rye, oats, millet, sorghum) and in vegetative structures such as tubers (modified stems, e.g., potatoes) and true roots (cassava, taro), and these plant parts are the most frequent sources of starch commercially.

Molecular Structure of Starch

Before looking at the synthesis of starch in amyloplasts and the structure of subsequent mature starch granules it is worthwhile to start by looking at the composition and configuration of the starch macromolecules. What will be described as "normal" starch is composed of two **polymers**, amylose and amylopectin, that occur in an approximate 1:3 ratio.

Amylose is an essentially linear polymer of α-D-glucopyranosyl units joined by (1→4) glycosidic bonds (Figure 7.1). There is evidence of a small degree of (1→6) branching in amylose, but the branching is infrequent and the branches are long, so its general physicochemical behavior conforms to the behavior of linear polymers. Amylose is the smaller of the two polymers: molecular weight is reported to be in the order of 10^4 to 10^5 (**degree of polymerization** [DP] 250–1000 D-glucose units). Its linear nature and lack of substitution allow it to strongly self-aggregate via H-bonds, and thus amylose is insoluble in cold water.

Amylopectin is highly branched and is also composed of α-D-glucopyranosyl **monomers**. Its linear chains are also joined by (1→4) glycosidic bonds. About 5% of the α-D-glucopyranosyl units also have another linear chain linked (1→6) to form the branch point (Figure 7.1). Each amylopectin molecule has only one reducing end. Amylopectin is much larger than amylose, *"truly a huge molecule, one of the largest found in nature"* (Delcour and Hoseney 2010), with molecular weight in the order of 10^6 to 10^8 corresponding to a DP of around 5000 to 50,000 D-glucose units. Amylopectin is considered by some to be too large to be truly soluble, but instead forms hydrated colloidal suspensions even after complete molecular dispersion in excess water under conditions of high shear and at temperatures exceeding 100°C. Smaller structural analogues of amylopectin (e.g., glycogen) are soluble in water, the branched nature slowing their capacity to self-associate. The generally accepted model for amylopectin structure is the cluster model shown in Figure 7.1. Linear amylopectin terminal chains longer than around 10 glucose units can adopt an ordered, crystalline, double-helical conformation (Copeland et al. 2009), and this is the basis of the crystallinity found in starch granules.

Amylose molecules can also form double helices. These strongly associate as a result of the linear nature of amylose and are more evident in starch that has been cooked and subsequently cooled. The α-glucan chains can also form a single helix with a hydrophobic core. This allows amylose in particular to complex with iodine and with free fatty acids (FFA). These attributes are important in starch analysis and functionality, respectively.

(A)   Amylopectin   Amylose

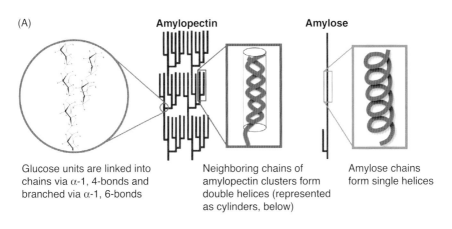

Glucose units are linked into   Neighboring chains of   Amylose chains
chains via α-1, 4-bonds and   amylopectin clusters form   form single helices
branched via α-1, 6-bonds   double helices (represented
                                    as cylinders, below)

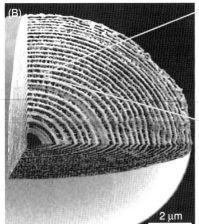

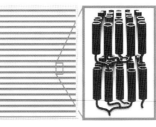

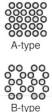

A-type

B-type

Alternating   Crystalline   Alternative
crystalline and   lamellae   arrangement of
amorphous   composed of   double helices
lamellae, repeated   aligned double   in crystalline
with 9-nm   helices of   lamellae (top
periodicity   amylopectin   view). Mixtures
                     (B-type shown)   of A and B are
                                          designated
Internal growth-ring structure of a starch granule   C-type
(adjusted composite image)

 Zeeman SC, et al. 2010.
Annu. Rev. Plant. Biol. 61:209–34

Figure 7.1   The composition and structure of starch granules. (a) A schematic representation of amylose and amylopectin, and the structures adopted by the constituent chains. (b) The relationship between the starch granule (composite image of potato granules, left) and amylopectin structure. Crystalline and amorphous lamellae arrange. . . to form the growth rings. (Reproduced from Zeeman et al. 2010 with permission.)

Notably, FFA complexed single-helix amylose does not form crystalline arrays and thus does not participate in starch recrystallization.

There are naturally occurring variations on this basic amylose-to-amylopectin (AM:AP) ratio of normal starch. For example, "waxy" starch has zero or almost zero amylose. (*Note: Waxy starch has nothing to do with the chemical entities "waxes," which are commonly esters of long-chain fatty alcohols or acids.*) Waxy starches are favored in chemical modifications as the absence of amylose reduces their tendency to recrystallize after being cooked into a molecular dispersion and subsequently cooled. Thus, waxy starches have less potential for **syneresis** (weeping) and shrinkage on cooling. Amylose, being linear, has a very high propensity for recrystallization. High-amylose starches are favored in film-forming applications and for the creation of one form of **resistant starch** (see the "Resistant starch" section below). Waxy and high-amylose variants can be found in wheat, barley, maize, and rice, among others.

Even small changes in AM:AP ratio can have profound effects on food texture. The Udon noodles of Japan are highly prized when they have a texture that combines softness with a distinctly elastic response. When selecting wheat for production of Udon flour, it is now known that it is advantageous to pick wheat with a slightly lower AM:AP ratio along with some other nonstarch traits. Favored wheat varieties have starch with a greater swelling power or hot paste viscosity. These attributes are related to the increase in the proportion of the highly branched amylopectin that swells on cooking and leads to the softer texture compared with Udon made from wheat flour with normal starch. It was later found that in wheat the loss of one copy of the gene coding for the enzyme "granule-bound starch synthase" (GBSS) leads to the small but significant decrease in amylose content. This phenomenon was only possible as a result of the hexaploid nature of wheat, in which three homologous copies of most genes are retained. In a diploid crop, such as barley or rice, loss of the single available copy of the GBSS gene leads to the loss of amylose synthesis and an effectively 100% amylopectin, or waxy, starch.

In practice, one can differentiate between amylose and amylopectin on the basis of the color formed when they are complexed with iodine (from an I_2/KI solution). (Refer to Unit 8, Microscopic Examination of Starch, in the Appendix.) The hue of the complex is very sensitive to the length of the starch chain. At the minimum, around the length of the amylopectin terminal chains, the complex is a reddish color ($\lambda_{max} =$ 530–550 nm). With chain lengths of the order of amylose, the color is a deep blue ($\lambda_{max} = 640$–660 nm) as a result of the inclusion of long arrays of poly(I_3). This phenomenon can be used to identify the presence of starch (e.g., in forensics). Differences in hue can give an indication of proportion of amylose and amylopectin. We use this simple color test in practice in barley breeding to confirm that our genetic markers for the "waxy" gene are correctly identifying our waxy breeding lines. There are quantitative methods for determining amylose content, including potentiometric assay of the starch/iodine complex. Another assay uses a defatted starch that is fully solubilized in an H-bond disruptor (commonly dimethyl sulfoxide; DMSO). Amylopectin is precipitated with the **lectin** concanavilin-A (Con-A; Gibson et al. 1997) and separated by centrifugation. Amylose remains dissolved in the DMSO supernatant and is subsequently hydrolyzed fully to glucose, which is assayed quantitatively and expressed as a proportion of the total starch in the sample. Summaries of other methods that can be used for amylose determination can be found in a number of excellent reviews (e.g., Gunaratne and Corke 2004; Zhu et al. 2008).

Starch Granules

Starch granules are fascinating. The way that starch is packaged in nature is completely unique among all polysaccharides. Starch exists in small, dense, discrete packages called granules that are insoluble in cold water and are energy storage organelles in plants. Granule dry weight is almost all glucose, giving them a high-energy density per unit volume. Starch granules do contain minor components, although there is contention that some molecules identified as minor components are actually contaminants from the plant matrix co-extracted during the isolation process (Delcour and Hoseney 2010). Lipids can comprise up to 1.5% of the dry weight of starch granules (Delcour et al. 2010). In cereal starches, these are mainly phospholipids and fatty acids. There are only trace amounts of lipids in tuber starches (Morrison 1998). Proteins associated with starch granules are mostly starch synthetic enzymes and surface proteins. The most important mineral component is phosphorous. It is a component of the phospholipids in cereal starches and occurs as phosphate mono-esters in potato starch. In the potato starch case, both the disruption of the starch molecular structure by the phosphate mono-ester groups and the electrostatic repulsion they provide as a result of a partial negative charge enhance the swelling characteristics of potato starch on cooking in water.

The physical density of starch granules is in the order of 1.5 g/cm^3, and this attribute in concert with their cold water insolubility makes them easy to separate from other plant tissues. Most of us who have cut a potato or other starch-rich plant material in water have seen the hard-to-move, almost pure white material in the bottom of the sink; this is granular starch, which, because its density is higher than water, sinks in a water column. The large molecular size of the starch polymers confers a low osmotic pressure on the granules. It is this attribute that allows plants to store such a high concentration of glucose with minimal effects on the water relations in their cells. Granules vary in size and shape among different plants: from 2 μm in high-amylose maize to 100 μm for potato. Wheat starch granules are lentil shaped and come in two size distributions with median diameters of around 10 and 50 μm. Starch granule shapes and sizes vary greatly. Larger granules tend to swell more on cooking, and hence granule size is a factor to be considered in starch functionality.

The internal structure of a starch granule is ordered. In this account, we will focus on normal starch as the model for discussion. Granular internal structure is concentrically layered (Figure 7.1). The layers, in their simplest conception, are made up of repeating layers, or growth rings, of crystallized terminal chains of amylopectin interlaid with amorphous regions of amylose and amylopectin branch points (Figure 7.1). There is a counterintuitive aspect to this structure. As noted above, after starch is cooked and allowed to cool, it is the linear amylose that has the higher propensity to crystallize, in accordance with generally held concepts of polymer crystallization. However, the timing of amylose synthesis in the **amyloplast** has it laid down in the same region as the AP branch points, simple steric hinderance effectively blocking amylose's ability to crystallize.

The semicrystalline nature of starch granule structure also allows granules to be investigated by techniques sensitive to the extent of crystallinity within a material: x-ray or electron diffraction, NMR, or simply polarized light micropscopy. In the latter, starch granules exhibit birefringence, or double refraction, seen as a cross-like image within the granule. (Refer to Unit 8, Microscopic Examination of Starch, in the Appendix.) Loss of birefringence is one of the indicators of starch gelatinization.

Gelatinization and Pasting: The Cooking of Starch

We do not eat a lot of raw starch, although we might want to consider eating more as raw granular starch is one form of **resistant starch**, which is considered to be protective against colorectal cancers. Generally, we cook starchy foods before consumption and almost always in the presence of water. Heating starch in the presence of water leads to an irreversible disruption of the granule structure at a temperature (the gelatinization temperature) that is characteristic of the botanical origin and molecular composition of the starch. If done for a sufficient length of time in the presence of excess water (water:starch ratio >1.5; Colonna and Buleon 2010), this can lead to a complete disruption of the granules and the creation of a molecular dispersion of starch molecules, a paste. Gelatinization and pasting have been described as a collection of events rather than single phenomenon (Atwell et al. 1988; Parker and Ring 2001). On heating, the internal ordered structure is disrupted and irreversibly destroyed. Concurrent with this, the granules swell to many times their original size, and amylose, the smaller of the two starch macromolecules, is solubilized and leaches from granules. The leaching may be driven by thermodynamic incompatibility (immiscibility) between amylose and amylopectin (Kalichevsky and Ring 1987). The basic molecular phenomena driving the disruption are the relaxation of the amorphous regions from their hard, vitreous, glassy state, to a malleable rubbery state characterized by increased molecular mobility and free volume due the combination of increased temperature and plasticization by water. There is a hypothesis that the structural relaxation is not a single relaxation but a series of them related to the specific characteristics of each glassy region (Biliaderis 2009), and this may be the reason that the glass transition is not resolved on standard thermograms (see "Differential scanning calorimetry" below). Concurrently, the thermally induced increase in molecular vibration literally shakes loose the H-bonds that stabilize the crystalline AP structure, allowing water to H-bond with the amylopectin, hydrating it, and pushing the amylopectin chains apart to create the swelling we observe. As heating is continued beyond the gelatinization temperature, the granules continue to swell, and a paste is formed. In the absence of shear, granules can still be seen as granule "ghosts" in photomicrographs, indicating that they retain their identity as discreet particles even after they are greatly swollen. Under conditions of shear, the granules are eventually completely disrupted, and a molecular dispersion is formed. The swelling of the granules can be detected macroscopically as an increase in the viscosity and volume of the granular suspension as it transitions to becoming a paste. Even under conditions of shear, especially early in the process, the major contributor to viscosity is the swollen granules (at temperatures higher than $\sim 60°C$, where amylose aggregation is not a factor; Parker and Ring 2001). As starch pastes are **thixotropic**, the paste viscosity decreases as shearing (stirring) continues at elevated temperatures. The thixotropic behavior, a time-dependent decrease in viscosity at a constant shear rate, results from granule disruption (no more particulate contribution to viscosity) and the alignment of macromolecules in the direction of the shearing force. The onset of pasting (a detectable viscosity increase at the "pasting temperature") and subsequent viscosity maximum and minimum at 95°C are diagnostic of starch composition within and between species (Figure 7.2) and are the bases of the Brabender Visco-Amylograph and Rapid Visco Analyzer (RVA; Crosbie and Ross 2007) tests for starch (See "Monitoring starch transitions" below). Under conditions of low or zero shear, granules remain somewhat intact, although swollen, and this

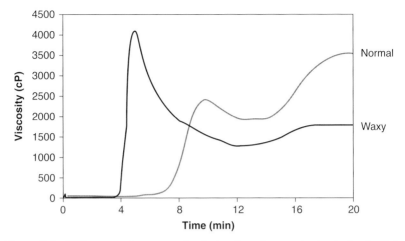

Figure 7.2   RVA curves for waxy wheat starch and normal wheat starch. Note the higher and more rapidly attained peak (hot paste) viscosity of the waxy starch resulting from its much more swollen granules. Note also the higher final viscosity of the normal wheat starch resulting from the incipient reassociation of the linear amylose fraction that is missing from the waxy starch. (Reproduced from Batey 2007 with permission.)

can be detected macroscopically as an increase in volume. This aspect is most easily experienced by comparing the volume of a cooked strand of noodle or spaghetti to its original volume, especially if one starts with dried noodles or pasta. The extent of swelling is largely related to the AM:AP ratio (higher proportions of AP leading to increased swelling, all other things remaining equal). This phenomenon is exploited in "swelling power" tests, and these are used routinely, for example, in screening potential new wheat varieties for their suitability for use in the Udon noodles described above. In limited-water food systems where granules are swollen but not disrupted, the deformability of the remnant granules greatly influences texture. In noodles, this is felt as increased softness as amylopectin content increases. The greater swelling power related to increased amylopectin concentration leads to more deformable granules, which are perceived as softer texture when the noodles are consumed (Ross et al. 1997).

Retrogradation and Gelation: The Cooling of Cooked Starch

The cooling of gelatinized starch leads to increased viscosity of pastes after the shear-induced minimum at high temperature (Figure 7.3: setback

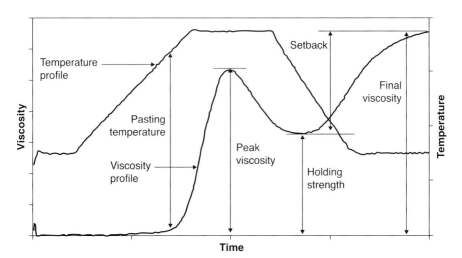

Figure 7.3   Typical complete RVA curve, showing the main parameters used to describe pasting as well as a typical time/temperature profile. See text for explanation of the main terms. (Reproduced from Batey 2007, with permission.)

and final viscosity). The viscosity increase is the first macroscopic expression of molecular level changes occurring in the gelatinized starch from an amorphous to a more ordered state (but not the ordered state of the granule). The change toward a more ordered state is termed **retrogradation**. The paste viscosity increase on cooling is related to the insolubility of amylose in cold water, and its tendency to aggregate as the H-bonds between amylose molecules become less transient (more persistent) as the temperature declines. In pure amylose solutions, as temperature drops the amylose rapidly becomes insoluble, precipitating at lower concentrations or forming a three-dimensional gel network at higher concentrations or at higher amylose molecular weights. Amylopectin molecules must also begin to self-associate, but the rate of aggregation or recrystallization is very much slower and of no functional relevance over short time frames. Based on the kinetics of recrystallization of amylose (fast and strong) and amylopectin (slow and weak), there is a two-stage process during the gelling of normal starches. First, amylose forms junction zones that support the structure of the incipient and initially weak gel; this happens over a period of seconds to minutes. The second stage is the gradual firming of the gel as the terminal amylopectin chains reassociate over a period of hours to days. The most common experience of the second-stage phenomenon we see in the western world is the firming of bread crumbs during storage, which is primarily a starch-related event. The importance of the amylose network in stabilizing gel structure can be observed by cooking up a waxy starch and seeing that it does not form a self-supporting gel immediately on cooling. Waxy starches can form gels, but they need high solid concentrations, to be held at low temperatures, may take a long time to set, and are weak (Kitamura 1996). Amylopectins with higher proportion of longer terminal chains also have a higher propensity to retrograde (Shi and Seib 1992). Rigidity or firmness of normal starch gels and crystallinity do not always increase together. It is possible to observe increasing gel rigidity in the absence of changes in x-ray diffraction patterns. This is thought to be a result of the creation of junction zones that are too small to be resolved, but are the nodes that help form the three-dimensional polymer network supporting the gel structure.

[Note: A word on terminology here: GELATINIZATION, a six-syllable word, has a very specific meaning as described above related to the behavior of raw starch as it is heated in water. GELATION, a three-syllable word, in the context of starch refers ONLY to the formation of a gel on the cooling of previously cooked (gelatinized) starch. Gelation is a general phenomenon of linear polymers when the solvent favors polymer/polymer associations over polymer/solvent associations (such as on the cooling of a suspension of gelatinized starch). Gelatinization is a phenomenon unique to starch.]

Of functional and nutritional importance are the differences in behavior of the retrograded starch molecules on reheating of a retrograded starch gel or starch-rich food such as bread or pasta. On reheating, it is possible to unwind amylopectin double helices. Familiar in the western world as the refreshing of bread by warming (even if the oven is set higher than $100°C$, the interior of moist bread does not exceed that temperature), the decrease in rigidity that is observed is the manifestation of this molecular level event. Amylose crystallites, by nature of their strong and stable associations, are not dissociated by heating to temperatures less than $150°C$ (Delcour and Hoseney 2010). If amylose crystallites did dissociate at temperatures below $100°C$, bread, for example, might collapse on reheating as it relies on the amylose network for much of its structure. One of the outcomes of reheating is an increase in overall molecular mobility. This means that the polymers may realign and increase the length of the junction zones on recooling. Exploiting this phenomenon in starch leads to a "perfection" of the crystalline linear α-glucan junction zones. With amylose, this can get to a point where the aggregates becomes resistant to enzymatic and acid hydrolysis, leading to the formation of RSIII resistant starch (See "Resistant starch" below). In normal starches, this phenomenon can be exploited or misused to speed up bread staling (crumb firming). The firming rate is maximized by cycling the bread through temperatures between $4°C$ and $25°C$ to maximize in turn the nucleation and propagation of the (primarily amylopectin) crystallites (Gray and BeMiller 2003). This behavior has been exploited to maximize bread staling rates for the production of products such as croutons (Slade et al. 1987). Keeping bread in the refrigerator, bringing out into the room temperature kitchen, and returning it to the refrigerator repeatedly accomplishes the same thing, though perhaps not to our advantage.

Monitoring Starch Transitions

As starch comes packed inside structurally complex packages (granules) and is a multicomponent entity, it stands to reason that the transitions of starch are also complex and may be monitored by a number of complementary methods (Parker and Ring 2001; Copeland et al. 2009). Commonly used methods range from empirical to fundamental and include microscopy, viscometric and other rheological methods, differential scanning calorimetry (DSC) and x-ray diffraction. Some of these are better suited to excess water systems (viscometry), and others are adaptable to limited water systems (water:starch ratio <1; Colonna and Buleon 2010).

Microscopy

Microscopes fitted with a polarizing filter and a "hot-stage" (controllable heating unit) can monitor gelatinization in excess water by observing the loss of birefringence as temperature increases. This technique is a valuable way to observe how the gelatinization temperature of an individual granule can be identified as the temperature of the first loss of birefringence and how the complete loss of birefringence occurs over a narrow temperature range of about 1°C. It is also easy to observe how the population of granules gelatinizes over a broader temperature range, up to 10°C. Optical microscopy is also a valuable tool for observing the variation in gelatinization temperatures of different types of starch (e.g., waxy vs. normal) within a species, and for starches across species and modifications.

Viscometric Methods

There are two common, largely empirical, viscometric methods in common use: the Viscoamylograph and the Rapid Visco-Analyzer (RVA). These instruments were specifically designed to measure viscosity of starch and flour pastes across a programmed time/temperature profile using a rotational sensor. A starch or flour slurry is cooked under controlled conditions, and the resistance against the rotating sensor is detected. The result is presented as a plot of apparent viscosity versus time (the "pasting curve"). There are numerous variants of solids concentrations, rotational speeds, and heating and cooling rates that can be applied for specific purposes (Ross and Bettge 2009). The pasting curves have several elements that are useful in characterizing a starch sample (Figure 7.3), and we have already encountered their use in diagnosing waxy versus nonwaxy starches (Figure 7.2).

Looking at Figure 7.3, moving left to right, the first parameter that can be observed is pasting temperature, seen as the first detectable increase in apparent viscosity. Pasting temperature is not the gelatinization temperature, which may better be observed as the first loss of birefringence using a hot-stage microscope. By the time we can see a detectable increase in paste viscosity, irreversible structural changes have occurred in starch granules (Batey 2007; Batey and Bason 2007), and so pasting is a consequence of gelatinization. Continued heating leads to a peak in the viscosity time curve (peak viscosity) that is related to the interaction of granule swelling and the break down of the swollen granules under shear. Continued mixing of the paste at 95°C leads to a decrease in viscosity related to granule disruption and thixotropic behavior of the paste as noted above (holding strength or minimum viscosity). On cooling, the paste again increases in viscosity. Called setback, the extent of this increase in viscosity is related often to starch amylose content. Starches with higher amylose content generally have higher setback or final viscosities as a result of the ability of amylose molecules to reassociate strongly on cooling. There are other more refined and fundamental viscometric methods that can be used, but these are not commonly used for starch characterization in food technology. Uses of the RVA for characterization of a variety of starches were reviewed in 2007 (Corke 2007; Rogers and Ross 2007).

Differential Scanning Calorimetry

DSC observes the difference in heat flow to a pan containing a sample versus a reference pan, commonly empty, as temperature increases or decreases at a predetermined constant rate (scanning). The differential heat flow allows us to observe a variety of exothermic and endothermic events in the sample. DSC is very useful for observing starch behavior across a range of water contents and allows us to observe behavior in the limited water conditions that are often typical of food formulations. A typical starch DSC thermogram of gelatinization of a cereal starch in excess water shows a large primary endothermic peak corresponding to the melting of the amylopectin crystallites (Figure 7.4). The onset temperature of the primary melting endotherm closely reflects the temperature of the first loss of

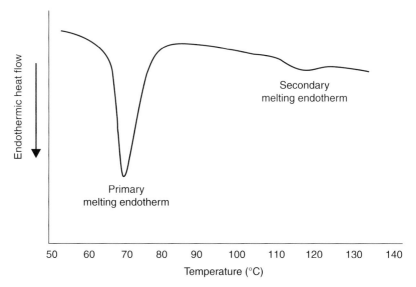

Figure 7.4 Schematic rendering of a typical differential scanning calorimetry output for a cereal starch heated in excess water. The primary endothermic peak is representative of amylopectin crystallite melting. The secondary peak is representative of the dissociation of amylose–lipid complexes.

birefringence. The melting endotherm shows a single well-resolved peak at high water content. As water content becomes limiting, the peaks shift to higher temperatures, showing how more energy is needed to gelatinize starch as there is less water available. In normal cereal starches, there is a secondary endotherm at around 110°C, corresponding to the dissociation of amylose/lipid complexes. DSC is also a useful tool to monitor the effect on starch of common food co-solutes, such as salt and sugar. It can also be used to monitor thermal transitions as gelatinized starch is cooled or reheated.

Starch Hydrolytic Enzymes

Starch hydrolytic enzymes are very important in the technology of starch. From the enzymes of liquefaction (viscosity reduction), to the enzymes of saccharification (fermentable sugar production), starch hydrolytic enzymes have been exploited for millennia (Samuel 1996, 2006). Ancient Egyptian bakers deliberately employed starch hydrolytic enzymes even if it was from a pragmatic foundation based on observable cause and effect: "*Preharvest or storage sprouting is precluded by the dry [Egyptian] climate.*

Deliberately germinated grain—that is, malt—must therefore have been used for some types of ancient Egyptian bread" (Samuel 1996). The technological outcomes of the uses of starch hydrolytic enzymes range from near to complete hydrolysis of starch for the production of fermentable sugars in the mash stage of brewing (or fuel ethanol production) to controlled and very limited hydrolysis for the production of fermentable sugars in bread making. In the bread case, the hydrolysis needs to be limited and controlled in order to leave sufficient unhydrolyzed starch to form the solid structure. Other limited hydrolyses are used to retard crumb firming in bread and to produce malto-dextrins for food uses.

α-*Amylase*

This family of enzymes is characterized by an endo (interior) hydrolysis pattern; they typically hydrolyze somewhat randomly, but within the molecule. Cereal α-amylases are specialized for α (1→4)bonds and have limited activity on α (1→6) bonds. The end products are a group of branched and linear malto-oligosaccharides, the smallest being maltose. The debranching enzymes in this group include pullulanase. This enzyme can catalyze hydrolysis of α (1→6) bonds. In technological application, the α-amylases can be used to rapidly decrease the viscosity of starch pastes. Viscosity reduction is particularly important in circumstances where high starch concentrations are used to maximize process yields. Rapid liquefaction by α-amylases decreases viscosity so that pumping and stirring operations are easier and so that heat diffusion rates can be maximized in the much less viscous liquefied substrate. Excess α-amylase produced in germination of wheat can degrade starch during bread making to such an extent that the bread structure can collapse, even in milder doses, leaving a sticky brownish and generally unpalatable crumb. Nonetheless α-amylases are almost essential in bread making. Very small doses of cereal or fungal α-amylases are almost universally used to ensure a slow but steady production of fermentable sugars for the fermentation organisms (yeasts, or yeast and bacteria in sourdoughs). Amylases attack the enzyme-susceptible damaged starch created during the dry milling of wheat flour. The very small doses are needed so that, in particular, the remaining amylose molecules are big enough to form a three-dimensional network that supports the bread structure on cooling.

The so-called maltogenic amylases, although part of the α-amylase family, are generally exo- in their hydrolysis pattern and conceptually remove the terminal chains of amylopectin. This pattern of hydrolysis leaves the amylopectin basically intact but with terminal chains too short to recrystallize, hence slowing or even stopping crumb firming during bread storage.

β-Amylase

β-Amylase is an exo-acting α-glucanase cleaving maltose from the nonreducing ends of the α-glucan chains. β-amylase also has limited ability to cleave α (1→6) bonds. Saccharification with β-amylase, even in concert with α-amylase, is therefore incomplete, and the end result is a mixture of maltose and branched and linear higher malto-oligosaccharides (malto-triose, malto-tetraose etc). β-Amylase and α-amylase act synergistically. Each cleavage of a starch chain by α-amylase exposes a new nonreducing end and another site for β-amylase attack. This is exploited in the mashing operation in brewing, in which the synergy speeds up the process. As β-amylase is present even in ungerminated cereals, the synergy is one reason that the dosing of α-amylase in bread flour needs to be done with great care to avoid excessive hydrolysis.

Modified Starches

There are many reasons to modify starches. Huber and BeMiller (2010) list a variety that include the following: inducing the ability to swell in cold water; reduced gelatinzation temperature; reduced energy to cook-up; altered adhesiveness characteristics; increased or decreased gel strength; improved freeze-thaw stability (reduced syneresis); processing tolerance (e.g., increased resistance of pastes to shear, acid, and heat); altered solubility profiles; and altered digestibility, among others. Many of the applications of modified starches go outside the boundaries of foods, including paper, pharmaceuticals, textiles, and novel biomaterials. Within foods applications include batters, coatings, and films, texturizing in dairy applications, fat mimetics, viscosity and clarity enhancement in sauces and fillings, cold water dispersibility ("instant starches"), viscosity and texture stabilization or enhancement in "extreme" environments: freezing, canning (high acid, shear, and heat), and emulsification (Table 7.1). Modifications can be categorized as physical or chemical, and in the modern era, biotechnological. Chemical modifications can be further categorized as converted (depolymerized), substituted (or stabilized), and cross-linked. Cross-linking and stabilizing reagents are generally applied to granular starches, and these still need to be cooked (gelatinized) to express their full functionality. Finally, the choice of the native starch to be modified (botanical source, AM:AP ratio, e.g., waxy, normal, or high amylose, etc.)

Table 7.1   A Summary of Starch Modifications, Their Outcomes, and Some Applications

Type of Modification	Properties	Applications
Pregelatinized and cold-water swelling	Dispersible in cold water	Convenience foods (e.g., instant puddings)
Partial hydrolysis	Depolymerized, reduced viscosity, increased retrogradation	Confectionary, batters, coatings
Oxidation	Depolymerized, reduced viscosity, improved clarity, reduced syneresis, improved cling	Batters and breadings, other coatings
Stabilization/starch esters	Reduced gelatinization temperature, decreased retrogradation and therefore a reduced tendency to form gels, improved clarity	Refrigerated and frozen foods, emulsion stabilizers
Stabilization/starch ethers	Reduced gelatinization temperature, high viscosity, decreased retrogradation and good freeze-thaw stability, improved clarity	Many applications: gravies, dips, sauces, fruit pie fillings, and puddings
Cross-linking	Increased cooked granule rigidity and stability, tolerance to high shear and temperature and low pH	Viscosity builders in high temperature and high shear processing; e.g., canning and retort processing, applications that require viscosity control during pumping operations
Exploiting genetic variability	e.g., waxy, high amylose	High viscosity and potential for RS3 resistant starch respectively

Source: Adapted from Singh et al. 2007.

interacts with the chosen modification(s) to determine the final functional properties. It is also very common in practice for more than one modification to be applied to a starch to achieve the desired outcomes.

Physical Modifications

Heat/moisture treatment (HMT) and mechanical shearing are common ways to physically modify starch. The most extensive of the HMTs is to pregelatinize the starch. In this process, starch is heated above its gelatinization temperature in the presence of water. After gelatinization, to a point where there is no residual birefringence, the starch is dried and ground to a predetermined particle size. The uses of Pregelatinized starch is used where cold water swelling or thickening is required, as for some instant pudding–type products. A variant on this type of starch is "cold water swelling" starch. These starches are gelatinized but retain an identifiable granular character, and this can be obtained by heating in aqueous alcohol solutions. Their applications include spreads and fillings, in which they impart high viscosity and glossiness without cook-up. The mildest of the HMT processes is annealing. In this case, the starch is heated in the presence of excess water to a temperature below the gelatinization temperature. Adding water to granules plasticizes the amorphous regions so that they go through a glassy to rubbery transition so that some rearrangement of the crystallites can occur. The outcome is a starch with increased crystallinity through growth of existing double helices rather than formation of new ones. The increased crystallinity leads to increased gelatinization and pasting temperatures but with variable effects on paste viscosities dependent on starch source and level of annealing applied (Tester and Debon 2000; Jayakody and Hoover 2008). Annealing seems to provide possibilities for mimicking some chemical modifications. Reducing granule swelling by annealing provides some thermal (and maybe shear) tolerance that could make the annealed starch suitable for canning applications, potentially providing an alternative to chemically cross-linked starches. Annealing has also been shown to enhance slowly digestible and resistant starch contents while retaining granule structure (Jayakody and Hoover 2008).

Physical modifications can also include dry milling (e.g., ball milling). In extensive ball milling, the end point is a complete loss of granule structure and crystallinity and increased susceptibility to enzymatic hydrolysis. These starches have greatly reduced pasting temperatures, lower paste viscosities, and weaker gels, partly as a result of reduced polymer size. A less dramatic but profound influence of dry milling is seen in bread making. During dry milling of hard wheat to bread flour, a small proportion (2–10%) of the starch granules are damaged (scuffed, scratched, split, or cracked). Granule damage is much less evident in flour milled from soft wheat. The damaged granules lose their internal structure and become more susceptible to enzymatic hydrolysis, as with the ball milled starches. In bread making, as we encountered above, the addition of an α-amylase source is common, particularly where the bread formulation has no added sugar(s). In these "lean" doughs, the damaged granules are the primary (really the only) α-amylase substrate and source of fermentable sugars during dough fermentation, before the dough is cooked and all the granules gelatinize and become susceptible to amylase attack.

Chemical Modifications

Depolymerization

Depolymerization of starches can be achieved using acidic or enzymatic hydrolysis, or via hypochlorite oxidation. Of the strictly hydrolytic processes, it is most common to employ enzymatic hydrolysis as a result of its lower energy input needs and greater control over the process. The result of hydrolysis is the production of starch hydrolysis products (SHPs). These are termed maltodextrins or corn syrup solids, the latter being of higher conversion levels (see Chapter 2, "Sugar Composition of Foods"). The degree of conversion of starch to smaller molecules can be defined by the dextrose equivalent (DE). This is defined as the number of reducing groups compared with the number of reducing groups in an equal weight of glucose (dextrose). This is more formally described as the molecular weight of glucose divided by the number average molecular weight of the SHP. A DE of 100 is glucose as each molecule has a reducing group. A DE of 0 is starch; each very large molecule has only 1 reducing group. The average degree of polymerization (DP) can be estimated be dividing 100 by the DE (e.g., a SHP with DE 20 has an average DP of 5 glucose units). SHPs with DE < 20 are often called maltodextrins, with DE > 20 corn syrup solids. As DE increases, SHPs become more soluble, sweeter, provide lower solution viscosity, are more fermentable, and provide greater effects in any **colligative property,** such as boiling point increase and osmotic pressure. SHPs are used as bulking agents, fat mimetics, and to restrict sucrose crystallization in the formation of hard, glassy candies.

Depolymerization is a side effect of the oxidation of starches. Oxidation of starch is usually affected by adding hypochlorite ions to an alkaline starch slurry heated to a temperature below gelatinization. The outcome is a starch that has had some of its hydroxyl groups oxidized to carbonyl ($-CHO$) or carboxyl ($-COOH$) groups. Glycosidic bonds adjacent to the oxidized groups are cleaved. The functional outcome is a starch with lowered paste viscosity (depolymerization), improved clarity, a reduced tendency to retrograde, and an increased propensity to bind to other materials as a result of the reactive oxidized groups.

Stabilization

Stabilized starches are substituted with new side groups, such as acetate, phosphate, or octenylsuccinate, in order to provide properties that are improved compared with their native precursors. These new side groups replace hydroxyl groups. The substituents can be ethers (e.g., hydroxypropyl, hydroxymethyl) or esters (e.g., acetate, phosphate) and are added to granular starch using so-called monofunctional reagents (e.g., acetic an-hydride).

Stabilized starches have some unifying properties. These are then mod-ulated by the exact nature of the substituting group: charged or un-charged, hydrophobic, and whether it is ether or ester bonded. Starch ethers tend to be somewhat more stable when exposed to acidic or alkaline environments. The general properties include the following:

- Reduced gelatinization temperatures
- Higher paste viscosities
- Improved clarity
- Resistance to retrogradation on the cooling of previously cooked starch
- Improved freeze-thaw stability
- Reduced syneresis

Most of these attributes can be rationalized by thinking of the sub-stituents, which are all simply bigger than the −OH group they replaced, as "disruptive structures" (Coultate 2009). These disruptive structures in-hibit interchain associations that would have occurred in their absence. This is salient at both the cook-up and cool-down stages of processing. Once inserted into a starch granule, substituents prevent the same close packing of starch molecules that was possible before their insertion. As a result of a slightly increased amount of free volume in the granule, wa-ter is better able to hydrate the granule, and so gelatinization temperature is decreased. This is even more evident where the substituent is a mono-starch phosphate that carries a negative charge at low ionic strengths of co-solutes and where the electrostatic repulsion amplifies the effect of the substituent. This also leads to more expanded granules on cook-up that may lead to increased hot paste viscosity. This circumstance occurs nat-urally in potato starch, which has naturally occurring mono-starch phos-phate esters.

On cooling, the substituents restrict the inter-chain associations of ret-rogradation. This has the molecular effect of restricting the growth of junc-tion zones, which is observed macroscopically as improved freeze-thaw stability and a reduced tendency for syneresis.

Cross-linking

Cross-linking in granular starches is achieved by reacting –OH groups with di-functional reagents (e.g., sodium tri-meta phosphate) that lead to cross-links such as di-starch phosphate esters and di-starch adipate (using adipic acid anhydride).

The effects of cross-linking are variable and depend on the starch type being modified and depend greatly on the amount of cross-linking. At very low levels of cross-linking, starch hot-paste viscosities may be increased as a result of the improved structural stability of the expanded gelatinized granules. As cross-linking extent increases, granule swelling is progressively restricted, leading to lower hot-paste viscosities, but viscosities that do not break down under conditions of high shear, high temperature, or acidity. The cross-linking can be regulated to provide low viscosity on first cook-up in order to facilitate pumping at low viscosities, with a progressive increase in viscosity with time that reaches the target viscosity in the final product. Cross-linked starches find their greatest application in extreme conditions, such as those imposed in canning and retorting, especially where the conditions are acidic.

Multiple Modifications

In reality, modified starches rarely contain only a single type of modification. Rather they are more commonly subjected to more than one modification, for example, being both cross-linked and substituted (stabilized). The choice of starting material also has a profound influence: whether it is a waxy or a normal starch, or whether, for example, one chooses to modify maize or potato starch.

Resistant Starch

"Resistant starch (RS) refers to the portion of starch and starch products that resist digestion as they pass through the gastrointestinal tract" (Nugent 2005). RS is broadly defined in order to cover all the varying starch polymorphs that are not digested in the human digestive tract earlier than the colon, where they, like other fiber sources, are fermented by the colonic bacteria. The major types of resistant starch are listed in Table 7.2. These include uncooked starches as well as starches that have been cooked (gelatinized) and recooled (retrograded). RS1 is based on starch that is physically "hidden" by other plant anatomical structures from the digestive process. RS2 is certain raw starches that are not, or are only slowly, digested. RS2 becomes available for digestion if it has been cooked. RS3 is most commonly based on normal, or even preferably high-amylose, starch sources. The reason for the need for amylose in RS3 is the enhanced capability of amylose to quickly and strongly recrystallize (retrograde) on cooling of a cooked starch matrix. The strong tendency of amylose to recrystallize and the ability to grow and make the amylose crystallites more perfect during

Table 7.2   Categorization of Resistant Starches, Their Sources and Attributes. Adapted from Nugent 2005.

RS Category	Sources	Attributes
RS1	Whole grains or seeds or partly milled products	Physically occluded starch, physical barriers to digestion such as intact cell walls, seed coats, etc.
RS2	Raw (uncooked) starches, particularly uncooked potatoes and unripe bananas	Raw starches with molecular structures that resist amylase attack, become fully digestible on cooking
RS3	Cooked and recooled (retrograded) starches, especially effective with high amylose starches and/or repeated heating and cooling cycles	Retrograded amylose crystallites become sufficiently perfect that they are resistant to amylase attack
RS4	Modified starches, often cross-linked	Substitutions and crosslinks hinder amylase binding and reduce susceptibility to hydrolysis

Source: Adapted from Nugent 2005.

repeated heating and cooling cycles is exploited in the food industry in order to create sources of RS3 for addition to foods (Haynes et al. 2002). Pressure cooking high-amylose starch to around 140°C and then cooling it to around 70°C in a cyclical process leads to new nucleation and subsequent propagation of the amylose crystallites and maximizes the amount of RS3 in the system. The bland taste and white color of RS makes it an attractive option for admixture to foods. Although not functionally well differentiated from microcrystalline cellulose, the divergent fates of cellulose and RS in the lower gut make RS an attractive additive in "healthy" formulations (see next paragraph). RS4 is chemically modified starch, in which extensive cross-linking restricts the ability of amylases to digest it by interfering with the ability of amylases to bind to the starch.

The perceived health benefits of RS are realized in the colon, where as-yet-undigested starch is fermented by the lower-bowel bacteria. All colonic fermentations produce acetic acid in abundance, but it is the higher levels of butyric acid that arise from the RS fermentation (compared with the higher amount of propionic acid fermented from nonstarch fiber sources) that are believed to be the genesis of the protective effects of RS against the onset and proliferation of colorectal cancers. Butyric acid is believed to act as a cell growth regulator but also contributes to other more general factors that improve bowel function, such as lowered fecal pH. In addition, as RS appears to be fermented in the distal (descending) colon, as opposed to nonstarch fibers that are fermented in the ascending and transverse colons, it extends these beneficial attributes further along the digestive tract.

Concluding Remarks

Starch is a crucial part of feeding the world. Accordingly, for future food scientists, a working knowledge of its sources, composition, variations, and responses to thermal processing are all key elements in a well-rounded knowledge base in the discipline. This pertains whether individuals are interested in food processing at its most modern and intricate or are primarily interested in foods subjected to minimal processing. Research showing the potential health benefits of resistant starches provides a further impetus for food science professionals to acquire the knowledge and skills to select the right starch for an application and to be able to manipulate it through processing in order to come up with the desired outcome, be it for processing, culinary delight, health benefit, or all three.

Another perspective that suggests a rich intellectual experience in studying starch is how observing starch and its transformations can illuminate many fundamental physicochemical phenomena. This pertains even in a "simple" product like bread made only with flour, water, salt, and yeast. The thermal transitions of starch provide ways to understand glass transitions, crystal melting, thixotropy, and shear thinning of polymer sols in stirred or pumped systems, recrystallization, gelation and its control, and the delicate interplay between substrate and enzyme: viewing starch as a variably vulnerable substrate for the host of amylases ready to attack it and fragment it all the way to glucose if possible. Students who become fascinated by seemingly endless possibilities for learning thrown up by this apparently simple polymer (it is just glucose after all) are encouraged to read further in a host of monographs and reviews on the subject (e.g., Thomas and Atwell 1997; Hancock and Tarbet 2000; BeMiller and Whistler 2009; Copeland et al. 2009; Bertolini 2010; Delcour et al. 2010; Delcour and Hoseney 2010; Zeeman et al. 2010).

Vocabulary

Amylase enzymes:
  α-amylase—Hydrolyzes α-1-4 glucosidic linkages with an endo mechanism
  β-amylase—Hydrolyzes α-1-4 glucosidic linkages with an exo mechanism

Glucoamylase (amyloglucosidase)—Hydrolyzes α-1-4 and α-1-6 glucosidic linkages in an exo-fashion from the nonreducing end
Pullulanase—Microbial enzyme specific for an α-1-6 glucosidic linkage
Birefringence—Characteristic cross-striations exhibited by starch granules when viewed under polarized light microscopy; evidence of crys-
tallinity
Corn syrup solids—Starch hydrolysate products with a DE > 20
Degree of polymerization (DP)—Number of sugar units in a polysaccharide molecule
Dextrose equivalent (DE)—Percentage of reducing groups in a starch hydrolysate on a dry weight basis
Gel—A colloidal system (solid/liquid dispersion) that exhibits properties
of both solids and liquids
Gelation—Formation of a gel; a general phenomenon exhibited by hydrocolloids
Gelatinization—Loss of a starch granule's internal structure due to water absorption at elevated temperatures in most food systems; a phenomenon unique to starch
Gelatinization temperature—Temperature at which an aqueous starch suspension loses its internal structure as evidenced by loss of birefringence
Glassy state—A solid state in which the molecules are not arranged in any
regular order; also known as the vitreous state
Glass transition—The reversible transition in the amorphous regions
within semicrystalline materials from a hard and brittle state into a
molten or rubber-like state
Maltodextrins—Starch hydrolysate products with a DE < 20
Pasting temperature—The temperature where an increase in viscosity of a
starch slurry is detectable on a starch viscometric instrument; pasting is
a consequence of starch gelatinization
Resistant starch—The portion of starch and starch products that resist digestion as they pass through the gastrointestinal tract
Setback—Increase in viscosity observed with a starch viscometric instrument when the starch paste is cooled
Starch retrogradation—Reaction that takes place in gelatinized starch in
which amylose and amylopectin chains realign themselves, becoming
irreversibly insoluble and causing gelation
Syneresis—The exuding of trapped water from a gel
Thixotropic—Fluids that display a decrease in viscosity over time at a constant shear rate
Waxy starch—Starch granules containing zero or negligible amylose
molecules

References

Atwell WA, Hood LF, Lineback DR, Varriano-Marston E, Zobel HF. 1988. The terminology and methodology associated with basic starch phenomena. *Cereal Food World* 33:306–311.

Batey IL. 2007. Interpretation of RVA curves. In: Crosbie GB, Ross AS, editors. *The RVA handbook*. St. Paul, MN: AACC-International Press, pp. 19–30.

Batey IL, Bason ML. 2007. Appendix 2: Definitions. In: Crosbie GB, Ross AS, editors. *The RVA handbook*. St. Paul, MN: AACC-International Press, pp. 136–140.

BeMiller J, Whistler R. 2009. *Starch chemistry and technology, 3rd ed*. Burlington, MA: Academic Press.

Bertolini A. 2010. *Starches: characterization, properties, and applications*. Boca Raton, FL: CRC Press.

Biliaderis CG. 2009. Structural transitions and related physical properties of starch. In: BeMiller J, Whistler R, editors. *Starch chemistry and technology, 3rd ed*. Burlington, MA: Academic Press, pp. 293–372.

Colonna P, Buleon A. 2010. Thermal transitions of starch. In: Bertolini A, editor. *Starches: characterization, properties, and applications*. Boca Raton, FL: CRC Press, pp. 71–102.

Copeland L, Blazek J, Salman H, Tang, MC. 2009. Form and functionality of starch. *Food Hydrocolloid* 23:1527–1534.

Corke H. 2007. Specialty cereal and noncereal starches. In: Crosbie GB, Ross AS, editors. *The RVA handbook*. St. Paul, MN: AACC-International Press, pp. 49–62.

Coultate T. 2009. *Food: the chemistry of its components*, 5th ed. Cambridge: RSC Publishing.

Crosbie GB, Ross AS. 2007. *The RVA handbook*. St. Paul, MN: AACC-International Press.

Delcour JA, Bruneel C, Derde LJ, Gomand SV, Pareyt B, Putseys JA, Wilderjans E, Lamberts L. 2010. Fate of starch in food processing: from raw materials to final food products. *Annu Rev Food Sci Technol* 1:87–111.

Delcour J, Hoseney RC. 2010. *Principles of cereal science and technology*. St. Paul, MN: AACC International Press.

Gibson TS, Solah VA, McCleary BVA. 1997. A procedure to measure amylose in cereal starches and flours with concanavalin A. *J Cereal Sci* 25:111–119.

Gray JA, BeMiller JN. 2003. Bread staling: molecular basis and control. *Comp Rev Food Sci/Food Safety* 2:1–21.

Gunaratne A, Corke H. 2004. Starch: analysis of quality. In: Corke H, Walker CE, Wrigley CW, editors. *Encyclopedia of grain science, Vol 3*. Oxford: Elsevier, pp. 202–212.

Hancock R, Tarbet B. 2000. The other double helix: the fascinating chemistry of starch. *J Chem Educ* 77:988–992.

Haynes L, Gimmler N, Locke JP III, Kweon M, Slade L, Levine H, inventors. 2002. Kraft Foods Holdings, Inc., assignee. March 5, 2002. *Enzyme-resistant starch for reduced-calorie flour replacer*. U.S. Patent 6,352,733.

Huber K, BeMiller JN. 2010. Modified starch. In: Bertolini A, editor. *Starches: characterization, properties, and applications*. Boca Raton, FL: CRC Press, pp. 145–203.

Jayakody L, Hoover R. 2008. Effect of annealing on the molecular structure and physicochemical properties of starches from different botanical origins: a review. *Carbohydr Polym* 74:691–703.

Kalichevsky MT, Ring SG. 1987. Incompatibility of amylose and amylopectin in aqueous-solution. *Carbohydr Res* 162:323–328.

Kitamura S. 1996. Starch polymers, natural and synthetic. In: Salamone JC, editor. *Polymeric materials encyclopedia*. Boca Raton, FL: CRC Press, pp. 7915–7922.

Morrison WR. 1998. Lipids in cereal starches: a review. *J Cereal Sci* 8:1–15.

Nugent AP. 2005. Health properties of resistant starch. British Nutrition Foundation, *Nutrition Bulletin* 30:27–54.

Parker R, Ring SG. 2001. Aspects of the physical chemistry of starch. *J Cereal Sci* 34:1–17.

Rogers R, Ross AS. 2007. Starch refining and modification applications. In: Crosbie GB, Ross AS, editors. *The RVA handbook*. St. Paul, MN: AACC-International Press, pp. 63–74.

Ross AS, Bettge AD. 2009. Passing the test on wheat end-use quality. In: Carver B, editor. *Wheat: science and trade*. New York: Wiley-Blackwell, pp. 455–493.

Ross AS, Crosbie GB, Quail KJ. 1997. Physicochemical properties of Australian flours influencing the texture of yellow alkaline noodles. *Cereal Chem* 74:814–820.

Samuel D. 1996. Investigation of ancient Egyptian baking and brewing methods by correlative microscopy. *Science* 273:488–490.

Samuel D. 2006. Modified starch. In: Torrence R, Barton H, editors. *Ancient starch research*. Walnut Creek, CA: Left Coast Press, pp. 205–216.

Shannon JC. 2009. Genetics and physiology of starch. In: BeMiller J, Whistler R, editors. *Starch chemistry and technology*, 3rd ed. Burlington, MA: Academic Press, pp. 23–82.

Shi Y-C, Seib PA. 1992. The structure of four waxy starches related to gelatinization and retrogradation. *Carbohyd Res* 227:131–145.

Singh J, Lovedeep K, McCarthy OJ. 2007. Factors influencing the physicochemical, morphological, thermal and rheological properties of some chemically modified starches for food applications: a review. *Food Hydrocolloids* 21:1–22.

Slade L, Oltzik R, Altomare RE, Medcalf DG, inventors. 1987. General Foods Corporation, assignee. April 14, 1987. *Accelerated staling of starch based products*. U.S. Patent 4,657,777.

Smith AM, Zeeman SC, Smith SM. 2005. Starch degradation. *Annu Rev Plant Biol* 56:73–98.

Tester RF, Debon SJJ. 2000. Annealing of starch: a review. *Int J Biol Macromol* 27:1–12.

Thomas DJ, Atwell WA. 1997. *Starches*. St. Paul, MN: Eagan Press.

Zeeman SC, Kossman J, Smith AM. 2010. Starch: its metabolism, evolution, and biotechnological modification in plants. *Annu Rev Plant Biol* 61:209–234.

Zhu T, Jackson DS, Wehling RL, Geera, B. 2008. Comparison of amylose determination methods and the development of a dual wavelength iodine binding technique. *Cereal Chem* 85:51–58.

8 Plant Cell Wall Polysaccharides

Bronwen G. Smith and Laurence D. Melton
Food Science Programme, The University of Auckland, Auckland, New Zealand

Introduction: Why Plant Cell Walls are Important

In plants, each cell has a wall surrounding its cell membrane. These walls have a complex composition and construction, which apparently differs from species to species and by function of the cell type. Each cell in a plant is joined to its neighbors through a region of the wall known as the middle lamella. The ability of plants to adopt a strong upright form, such as trees, or maintain delicate structures, such as fruit and flowers, is due to the presence of the cell wall.

Walls come in two mains types: primary or secondary. Primary cell walls are thin and deposited while the growing cell is expanding. This

Food Carbohydrate Chemistry, First Edition. Ronald E. Wrolstad.
© 2012 John Wiley & Sons, Inc. Published 2012 by John Wiley & Sons, Inc.

type is found in parenchyma tissue and constitutes most of what we eat in plant foods. In contrast, secondary walls are usually thicker and are deposited over the primary cell wall usually as cell expansion ceases in association with specialized development (e.g., the gritty cells, called scle-reids, in pears). In addition to providing mechanical support for the plant and determining structure and form, walls provide a barrier to the envi-ronment, and hence the walls of the epidermal cells may be impregnated with the hydrophobic substances cutin and suberin and have a waxy layer over them. The epidermal walls provide the first line of defense against pathogens.

Walls are composed of complex polysaccharides in varying amounts, including cellulose, xyloglucans, heteroxylans, mannans, and pectic polysaccharides. Plant cell walls from fruit, vegetables, and cereals are most of the dietary fiber in the diet of most people. Hence, plant cell walls are important for human health.

Note: Dietary fiber is regarded by Codex Alimentarius as polysaccharides in foods or extracted from foods or edible synthetic polysaccharides that have a beneficial physiological effect (e.g., decreases intestinal transit time, lowers LDL cholesterol). Whether oligosaccharides are to be included is to be decided by each national authority. This means plant cell walls, resistant starch, commercial pectins, alginate, agar, and carboxymethyl cellulose can be regarded as dietary fiber, but only in some countries would inulin and maltodextrin oligosaccharides of DP < 10 be allowed.

Because they provide structure to the plant, they also provide us with a range of textures, which is one of the attractions of plant foods. Walls are not static in terms of their composition. Moreover, because there is a diversity of form and function among plants as they respond to the drives of development, so there are corresponding changes in composition. For example, much of the softening in texture of fruits as they ripen can be attributed to changes in the structure of the pectic polysaccharides in the walls.

Texture also changes during cooking with interactions among the con-stituents altered, and some polysaccharides can be leached out of the walls. This is especially true for the pectic polysaccharides. Aside from the nutritional value of dietary fiber, cell walls impact human life in other ways. As they are a component of animal fodder, they form the basis of agricultural economies in the production of meat, wool, and dairy foods. Walls are also important industrially as wood, paper, and textiles such as cotton and linen. Dietary fibers from different sources can be used in the manufacture of foods (Harris and Smith 2006).

A wealth of information exists about the types of polysaccharides in the walls, but much less is known about the fine architecture or the way in which the polysaccharides are assembled to form this robust structure. Early scientific knowledge about cell walls came from microscopy of cork tissue by Robert Hooke in 1665. Since then, the focus has been on finding out about their composition using chemical analytical procedures. This means the walls must first be isolated from the plant material. This is usually followed by acid and/or enzymic treatments to break up the polysaccharides into their constituent monosaccharides followed by identification of the monosaccharides and inferred reconstruction of the polysaccharides.

Cellulose

Cellulose is the most abundant naturally occurring substance on earth. Each year plants form 1.5 trillion tons of cellulose. Cellulose is composed of glucose units linked β 1,4 to give the polysaccharide. Every second glucose is turned over, so that the repeating unit of the polymers is the disaccharide cellobiose (Figure 8.1). The individual glucan chains (Figure 8.2), which form the cellulose molecules, assemble together to form a microfibril (Figure 8.3). Cellulose molecules commonly consist of 300–1700 glucoses; however, in cotton fibers, they can have a degree of polymerization (DP) of 8000 or more. Whereas cellulose can make up one-third or more of the cell wall of many plants, walls of the endosperm of cereal grains often contain substantially less.

Historically, the microfibril was thought to be composed of 36 cellulose molecules held together by hydrogen bonds and van der Waals forces (Figure 8.3). Although this is true in some plants, in fruit and vegetables, microfibrils about 3 nm in diameter composed of around 25 molecules are

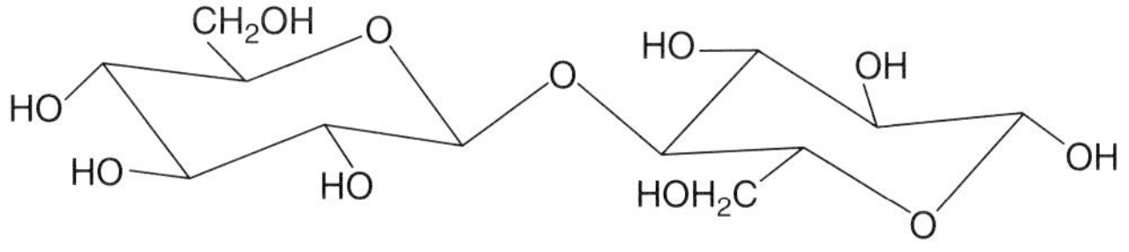

Figure 8.1   Cellobiose.

→4)-β-D-Glc*p*-(1→4)-β-D-Glc*p*-(1→4)-β-D-Glc*p*-(1→4)-β-D-Glc*p*-(1→4)-β-D-Glc*p*-(1→4)-β-D-Glc*p*-(1→

Figure 8.2   Cellulose molecule.

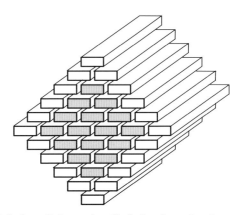

Figure 8.3   Model of a cellulose microfibril showing a 6 × 6 array.

commonly found (Newman et al. 1994; Smith et al. 1998). Cellulose microfibrils of larger diameter are known, but these are found in tougher material than food, such as wood and seaweed. As much as 96% of the mechanical strength of cell wall in fruit is attributed to the cellulose microfibrils. Therefore, it is perhaps not surprising that one of the smallest diameter microfibrils discovered so far is in strawberries (Koh et al. 1997). In summary, different plants can have microfibrils of different diameters, although in a specified plant only one size is apparent.

The cellulose microfibrils can be found adhering together as fibrils, for example, in cotton and other fibers. However, in *Arabidopsis*, the most studied of all plants, the cellulose exists as individual microfibrils (Davies and Harris 2003), and it seems likely that this will be the case for the soft edible tissue of vegetables and fruit. Atomic force microscopy shows that the cellulose microfibrils in celery cell walls are in ordered layers (Thimm et al. 2000).

The cellulose microfibrils are so orderly, and the three-dimensional interaction between them is so extensive that they are crystalline. Consequently, they are very difficult to break down. You would need to heat them in concentrated sulfuric acid all night if you wished to hydrolyze them down to glucose. In contrast, nature uses a battery of enzymes.

Because cellulose is so stable, it has been modified for use in the food industry by grinding to give a fine powder (microcrystalline cellulose) or by making chemical derivatives that are water soluble (methylcellulose, carboxymethylcellulose, hydroxypropylmethylcellulose). In the manufacture of these hydrocolloids, the first step is partial solubilization of cellulose by treatment with strong alkali, which is very effective at breaking hydrogen bonds. Disruption of the intermolecular hydrogen bonds between adjacent cellulose molecules allows for penetration of water. To produce methylcellulose, treatment with methyl chloride will form methyl ethers, the maximum degree of substitution (DS) being three methyl groups per sugar unit. The typical DS range for methylcellulose is 1.3–2.6. Water solubility occurs for steric reasons. The linear conformation of the cellulose molecule is altered by the nonpolar methoxy substituents, which prevent intermolecular association and allow for water penetration.

Carboxymethylcellulose is produced by reaction of alkali-solubilized cellulose with monochloroacetic acid. A maximum DS of 0.9 is allowed for use in foods. Carboxymethylcellulose is used in salad dressings and to prevent syneresis in pie fillings. Treating alkali-solubilized cellulose with propylene oxide forms hydroxylpropylcellulose. Hydroxylpropyl ethers are formed, and a DS > 3 per sugar unit is possible because the hydroxyl group of the hydroxylpropyl substituent can be derivatized. Solutions of hydroxylpropyl cellulose can form a gel at 60–80°C. An interesting application is its use in gluten-free flours to produce breads for people allergic to gluten. Cellulose whiskers formed by frayed cellulose molecules on the surface of microfibrils are frequently used in nanotechnology. For further information on cellulose, see Klemm et al. (2005).

Hemicelluloses

Xyloglucans

Xyloglucans are typically the most abundant hemicellulose in fruit and vegetables. The backbone is glucose-linked β-1,4 and three out of four glucoses have a xylose attached at the C-6 position (Figure 8.4). Galactose

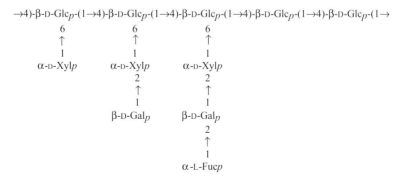

Figure 8.4   Xyloglucan.

and fucose and other sugars may also be present as part of the side chains, depending on the source. Xyloglucans play important roles in cell wall expansion in young plant tissues and in softening during fruit ripening.

Heteroxylans

Heteroxylans are based on a backbone of xylose residues linked together β-1,4. These xylose residues may be substituted with arabinose at the C-3 and sometimes C-2 position to form arabinoxylans (Figure 8.5).

Arabinoxylans are major components of the walls of endosperm cells of cereal grains, and hence become a component of flours from these grains following milling and are important in human nutrition. The xylose residues may also be substituted with glucuronic acid, or 4-O-methylglucuronic acid, usually at the C-2 position, as well as arabinose, to form glucuronoarabinoxylans (GAXs; Figure 8.6). The arabinose residues may be further substituted with ferulic acid. Polysaccharides with this composition are found in the primary unlignified cell walls of grasses and related plants. GAXs are therefore important components of some forages for animals, such as cows and sheep.

Figure 8.5   Arabinoxylan.

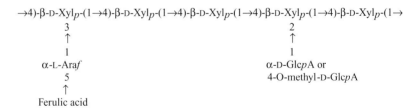

Figure 8.6   Glucuronoarabinoxylan (GAX).

(1→ 3),(1→ 4)-β-D-Glucans

(1→3),(1→4)-β-D-Glucans are only found in the cell walls of plants in the botanical order Poales, which includes the cereals and grasses (Smith and Harris 1999; Harris and Fincher 2009). These polysaccharides are found in the walls of the grains along with arabinoxylans, often making up a substantial proportion of the wall content. In the walls of vegetative tissue of growing grass, they are found together with cellulose and GAXs. Walls from oat and barley grains, in particular, are rich in (1→3),(1→4)-β-D-glucans. This is important for human health because these polysaccharides have some efficacy in lowering blood cholesterol levels (Wolever et al. 2010). (1→3),(1→4)-β-D-Glucans consist of glucose molecules linked together β-(1→4) with the regularity being interrupted every third or fourth glucose unit with a β-(1→3) linkage. This means that the shape of the polysaccharide is that of a long chain with bends provided by the change in the type of linkage from β-(1→4) to β-(1→3).

Mannans

Mannans occur in the primary walls of growing plants, usually as galactoglucomannans or sometimes glucomannans, but only in small amounts, where their role is structural. For a short review on mannans in growing plants, see Melton et al. (2009). Some occur as storage polysaccharides in the endosperm cell walls of seeds as mannans or galactomannans (for example, guar gum and locust bean gum). The mannan backbone consists of β-(1→4)–linked mannose units. In galactomannans, galactose is attached at the 6 position to some of the mannose residues. For more on storage galactomannans, see Gidley and Reid (2006).

Pectic Polysaccharides

Pectic polysaccharides that occur in plants are possibly the largest and most complex naturally occurring molecules in existence. The pectin used in the food industry is commonly extracted from apple pomace or orange waste, and as a result of processing is degraded to a simpler polysaccharide. Commercial pectin is largely a polymer of D-galacturonic acid linked α-1,4 (Figure 8.7), which contains small amounts of rhamnose.

The regions of the pectin molecules that are not esterified can cross-link to another pectin molecule via divalent charges on calcium ions. This arrangement is referred to as the "egg box model," with the calcium ions as the eggs within the junction zone (Grant et al. 1973) (Figure 8.8). This is one way pectins can interact. Another is by hydrophobic interactions of

→4)-α-D-Gal*p*A-(1→4)-α-D-Gal*p*A-(1→4)-α-D-Gal*p*A-(1→4)-α-D-Gal*p*A-(1→4)-α-D-Gal*p*A-(1→

Figure 8.7   Homogalacturonan.

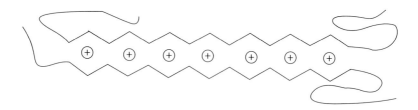

$\oplus$ denotes calcium ions cross linking adjacent regions of homogalacturonans

Figure 8.8 Egg box model.

methyl groups on adjacent pectins, which is what occurs when jams and preserves are made from fruit and sugar.

Pectins are described by their degree of methyl esterification (DE), which is the percentage of carboxyl groups esterified with methanol. If the percentage of methyl esterification is greater than 50%,it is called high-methoxyl pectin and if less than 50% it is called low-methoxyl pectin. In the food industry, knowledge of the degree of methylation and acetylation of pectins is useful because the extent of methylation and acetylation affects the gelling properties; the higher the degree of methylation, the greater the capacity to form gels, whereas a higher degree of acetylation inhibits gelling (BeMiller 2007). Different grades of high-methoxyl pectin are available commercially. "Rapid-set" pectin has a DE of 72–75% and sets within 20–70 sec at pH 3.0–3.1 (BeMiller 2007). This occurs with a pectin content of 0.3% and a sugar content of c.a. 65%. Rapid gelation allows for uniform distribution of suspended fruit throughout the gel, which is desirable for preserves and marmalades. "Medium-set" pectins have a DE of 68–71%, whereas "slow-set" pectins have DE values of 62–68%. Gelation temperatures vary from about 35°C to 90°C. The pH is critical for gel formation. "Slow-set" pectins require a lower pH than "rapid-set" pectins. Ionic repulsion of carboxylate anions will prevent inter- and intramolecular association of polygalacturonic acid chains and formation of junction zones. Increased acid (lower pH) is required to protonate carboxylate anions.

Pectin in the plant cell wall is far more complex. Pectic polysaccharides in plant cell walls generally consist of a backbone of galacturonic acid (GalA) residues interspersed with rhamnose residues to which are attached side chains of arabinans, galactans, and arabinogalactans (Caffall and Mohnen 2009; Figures 8.9 and 8.10).

Plants contain two general types of pectin-degrading enzymes: polygalacturonase, which hydrolyzes the α-1-4 glycosidic linkage between the

→4)-α-D-GalpA-(1→2)-α-L-Rha-(1→4)-α-D-GalpA-(1→2)-α-L-Rha-(1→4)-α-D-GalpA-(1→
                         4
                         ↑
                         1
                  sidechains of
        arabinans, galactans, arabinogalactans

Figure 8.9   Rhamnogalacturonans.

galacturonic acid units, and pectin methyl esterase, which hydrolyzes the methyl ester functional group. A combination of these enzymes plays an important role in the softening of fruit tissue during ripening. An interesting food application is improvement of the texture of frozen and canned vegetables (McFeeters 1985). A mild blanch at 50–80°C activates native pectin methyl esterase, which de-esterifies pectin and allows for calcium cross-linking of pectin molecules. A firmer texture results. The enzyme is subsequently inactivated by a high-temperature blanch or by heat processing. Pectic enzymes are used in juice processing to increase yield, and for fruit maceration, cloud stabilization, and juice clarification.

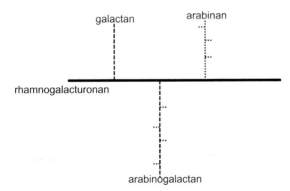

Figure 8.10   Schematic diagram showing sidechain attachments to rhamnogalac-turonans.

Interactions between Polysaccharides and Cellulose

As well as interactions of individual wall polysaccharides with each other, they can interact with different polysaccharides. Xyloglucan adheres to cellulose by hydrogen bonding. Hydrogen bonds are weak and easily broken, but if there are a large number of them, they are equivalent to a strong interaction as found in cellulose microfibrils. Extensive hydrogen bonding is possible because of the regular repeating structure of cellulose and the glucan backbone of xyloglucan, which in fact is a cellulose molecule. Likewise, the regular backbones of xylans and mannans allow them to bind to cellulose by hydrogen bonding (Melton et al. 2009). More recently, the regular repeating galactan and arabinan side chains of pectin have been shown to adhere to cellulose (Zykwinska et al. 2007). Such interactions help to explain how the polysaccharides are assembled to form a plant cell wall.

The Plant Cell Wall Structure

The three-dimensional structure of the cell wall has not been fully elucidated. Figure 8.11 shows a model of the primary cell wall as found in fruit. In this model, cellulose microfibrils are cross-linked by xyloglucan to make a three-dimensional network. Other polysaccharides, including galactan and arabinan side-chains of pectin, and mannans are attached to the surface of the cellulose microfibrils. A pectin gel network involving Ca^{2+} cross-links is another major network. The attachment of the galactan and arabinan side chains of pectin to cellulose results in a super-network. Approximately 5% (by dry weight) of the wall is structural protein, which constitutes a third network (not shown). Because each plant has different proportions of the polysaccharides in its walls, each will have different cell wall architecture. It is important to remember that water is the largest component of many walls, which is not surprising given the excellent water-holding capacity of polysaccharides.

In summary, although all cell walls consist of cellulose, hemicelluloses, and pectic polysaccharides, the proportions of each vary from plant to plant and with cell type. Because the polysaccharide composition varies, we can also expect the three-dimensional structure of the walls of plants to be different.

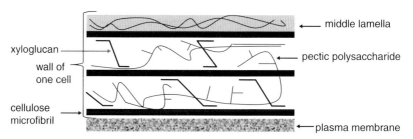

Figure 8.11   Schematic diagram of a primary cell wall from fruit.

Vocabulary

DE—Degree of esterification

DP—Degree of polymerization

DS—Degree of substitution; average number of substituted OH groups/ sugar unit

Gelation—Formation of a gel (A gel is a solid/liquid dispersion that exhibits properties of both solids and liquids.

References

BeMiller JN. 2007. *Carbohydrate chemistry for food scientists, 2nd ed.*, St Paul, MN: AACC International, pp. 303–311.

Caffall KH, Mohnen D. 2009. The structure, function, and biosynthesis of plant cell wall pectic polysaccharides. *Carbohydr Res* 344:1879–1900.

Davies LM, Harris PJ. 2003. Atomic force microscopy of microfibrils in primary cell walls. *Planta* 217:283–289.

Gidley MJ, Reid JSG. 2006. Galactomannans and other cell wall storage polysaccharides in seeds. In: Stephen AM, Phillips GO, Williams PA, editors. *Food polysaccharides and their applications.* Boca Raton, FL: CRC Press, pp. 181–215.

Grant GT, Morris ER, Rees DA, Smith PJC, Thom D. 1973. Biological interactions between polysaccharides and divalent cations: the egg-box model. *FEBS Lett* 32:195–198.

Harris PJ, Fincher GB. 2009. Distribution, fine structure and function of (1,3;1,4)-β-glucans in the grasses and other taxa. In: Bacic A, Fincher GB, Stone BA, editors. *Chemistry, biochemistry and biology of (1→3)-β-glucans and related polysaccharides.* San Diego, CA: Academic Press, Elsevier, pp. 621–654.

Harris PJ, Smith BG. 2006. Plant cell walls and cell-wall polysaccharides: structures properties and uses in food products. *Int J Food Sci Tech* 41(S2): 129–143.

Klemm D, Heublein B, Fink H-P, Bohn A. 2005. Cellulose: fascinating biopolymer and sustainable raw material. *Angew Chem Int Ed* 44:3358–3393.

Koh TH, Melton LD, Newman RH. 1997. Solid-state ^{13}C NMR characterization of cell walls of ripening strawberries. *Can J Bot* 75:1957–1964.

McFeeters RF. 1985. Changes in pectin and cellulose during processing. In: Richardson T, Finley JW, editors. *Chemical changes in food during processing.* Westport, CT: AVI Publishing, pp. 347–372.

Melton LD, Smith BG, Ibrahim R, Schroeder R. 2009. Mannans in primary and secondary plant cell walls. *N Z J Forest Sci* 39:153–160.

Newman RH, Ha M-A, Melton LD. 1994. Solid-state ^{13}C NMR investigation of molecular ordering in the cellulose of apple cell walls. *J Agric Food Chem* 42:1402–1408.

Smith BG, Harris PJ, Melton LD, Newman RH. 1998. Crystalline cellulose in hydrated primary cell walls of three monocotyledons and one dicotyledon. *Plant Cell Physiol* 39:711–720.

Smith BG, Harris PJ. 1999. Composition of the cell walls of the Poales. Primary cell walls of the Poaceae are not unique. *Biochem Syst Ecol* 27:33–53.

Thimm JC, Burritt DJ, Ducker WA, Melton LD. 2000. Celery (*Apium graveolens* L.) parenchyma cell walls examined by atomic force microscopy: effect of dehydration on cellulose microfibrils. *Planta* 212:25–32.

Wolever TMS, Tosh SM, Gibbs AL, Brand-Miller J, Duncan AM, Hart V, Lamarche B, Thomson BA, Duss R, Wood PJ. 2010. Physicochemical properties of oat β-glucan influence its ability to reduce serum LDL cholesterol in humans: a randomized clinical trial. *Am J Clin Nutr* 92:723–732.

Zykwinska A, Thibault J-F, Ralet M-C. 2007. Organization of pectic arabinan and galactan in association with cellulose microfibrils in primary cell walls and related models envisaged. *J Exp Bot* 58:1795–1802.

9 Nutritional Roles of Carbohydrates

Introduction

Carbohydrates receive a lot of bad press. It is an all too common misconception that dietary carbohydrates are problematic, and that humans would benefit from their reduced consumption. Although restrictions of certain carbohydrates are recommended for those individuals -with aberrant glucose metabolism, most humans will benefit from consuming substantial amounts of carbohydrate. Humans as a species evolved to use α-glucans as their principal energy source. Health organizations recommend that carbohydrates provide 55–60% of caloric intake (Whistler and BeMiller 1997). Worldwide, carbohydrates provide 70–80% of human

Food Carbohydrate Chemistry, First Edition. Ronald E. Wrolstad.
© 2012 John Wiley & Sons, Inc. Published 2012 by John Wiley & Sons, Inc.

caloric intake, whereas in the United States they supply less than that percentage, with widely varying amounts from individual to individual (BeMiller and Huber 2008). In addition, consumption of nondigestible carbohydrate is important for optimum health of the digestive tract. It is critical that food scientists have a rudimentary knowledge of carbohydrate nutrition so that they can appreciate the impact that processing, ingredients, product formulation, etc. can have on nutritional quality.

The intent of this chapter is to briefly describe the digestion, absorption, metabolism, and metabolic fate of dietary carbohydrates. Issues concerning carbohydrates and human disease will be presented along with recommendations for optimal nutritional well being. For more detailed information, the following sources are recommended: Khanna et al. 2006; Raben and Hermansen 2006; Gropper et al. 2009.

The Digestive Process: From the Bucchal Cavity through the Small Intestine

The digestion, assimilation, and transit of carbohydrates through the human digestive tract are summarized in Figure 9.1 as a flow sheet of unit operations. Particle size reduction occurs in the bucchal cavity (mouth) through chewing. Saliva, which is 99.5% water, dissolves some compounds and assists salivary mucoproteins in lubricating food. Starch hydrolysis is initiated by the action of salivary α-amylase on internal α-1-4-glucosidic bonds. Humans produce 1–2 liters of saliva per day, with saliva production ceasing during sleep, dehydration, and moments of anxiety or extreme mental effort. (Thus, the dry-mouth sensation that can occur during lectures can be caused by at least three different conditions.)

Action of salivary α-amylase with its pH optimum of 6.7–7.0 is greatly reduced by the low pH of the stomach. The pancreas adjusts the pH to neutrality by secretion of bicarbonate. Pancreatic α-amylase is also secreted with starch being hydrolyzed to oligosaccharides, dextrins, and maltose in the duodenum. The small intestine, which consists of the duodenum, jejunum, and ileum, is approximately 10 feet long. It is the major site for nutrient absorption. The folds, villi, and microvilli brush border of the small intestine combine to give a very large surface area of approximately 300 m^2. The glycocalyx is a surface coating of the microvilli that contains the following carbohydrase enzymes: maltase, isomaltase, lactase, sucrase, and trehalase. These enzymes are all α-glucosidases, except for lactase, which is a β-galactosidase. Only monosaccharides are absorbed, so oligosaccharides need to be hydrolyzed before absorption can take place.

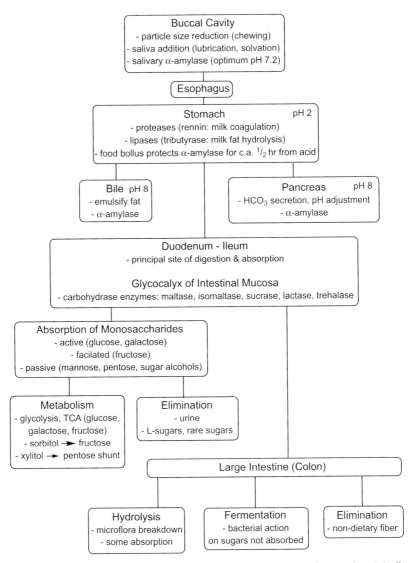

Figure 9.1   Flow diagram summarizing the digestion, absorption, and metabolic fate of carbohydrates.

Absorption of Sugars

Sugars differ as to their rate of absorption across the intestinal wall. In **active transport**, sugars will be absorbed against a concentration gradient. The process requires energy and a specific carrier, which is a

protein complex. The structural requirements for active sugar transport
are as follows:

- Pyranose structure with primary alcohol on C-5
- Equatorial OH on C-2
- C-1 conformation
- β-Configuration

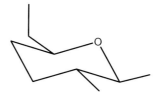

Structure 9.1

Thus, D-galactose and D-glucose meet the structural requirements for
active transport, mutarotation enabling the conformational and configurational prerequisites. Active transport makes for highly efficient absorption
and use of sugars. Galactose is actually absorbed at a slightly faster rate
than glucose. In **passive absorption**, sugar diffusion takes place according to differences in osmotic pressure. Absorption will not occur against a
concentration gradient. Rates of passive absorption are much slower than
those of active transport (e.g., ten times less than glucose; Zapsalis and
Beck 1985). Pentoses, L-sugars, sorbitol, xylitol, and hexoses other than
glucose, galactose, and fructose are passively absorbed. Fructose is absorbed by **facilitated transport**. The process proceeds down a concentration gradient. Because the liver very efficiently absorbs fructose, there is
essentially no circulating fructose in the bloodstream. Absorption of fructose is independent of glucose blood levels and much slower than that of
glucose.

Humans absorb glucose very efficiently. This is advantageous in obtain-
ing energy for body processes; however, it is problematic for people with
disorders such as diabetes, and may be a contributing factor to the issue of
obesity. Although glucose and sorbitol have the same caloric value, they
differ considerably as to their rate of absorption. The **glycemic index (GI)**
has been adopted as a measure of the effect of carbohydrate foods on blood
sugar levels. Subjects consume a test food containing 50 g of available carbohydrate, and the elevation of blood glucose over the base-line level is
measured over a 2-hour period and compared with a reference food, either glucose or white bread. Carbohydrate foods that break down quickly

during digestion and release glucose rapidly into the bloodstream have a high GI.

$$\text{Glycemic index} = \frac{\text{Blood glucose of test food}}{\text{Blood glucose of reference food}} \times 100 \qquad (9.2)$$

The concept was developed by David Jenkins and colleagues (1981) at the University of Toronto in their efforts to categorize foods as to which were best for people with diabetes. The GIs of some sugars and selected foods are shown in Table 9.1. It should be emphasized that large variations in GI occur between individuals (gender, age, type of diabetes) and measurement variables, for example, preprandial blood glucose level, eating time, food matrix, nature of starch, fat, protein and fiber content, and even food temperature. (A boiled red potato consumed hot has a GI of 89.4, whereas the same potato cold has a GI of 56.2; Gropper et al. 2009). GI tables are helpful to diabetics in making food choices and find application in diet plans such as the South Beach diet (Foster-Powell et al. 2002). Low-GI foods (<55) include most fruits and vegetables (except for potatoes and watermelon), 100% stone-ground whole wheat or pumpernickel bread, oatmeal, pasta, lima/butter beans, peas, milk, and yogurt (American Diabetes Association 2010). Whole wheat bread and basmati

Table 9.1   Glycemic Index of Common Foods[a]

Foods	GI (white bread)	GI (glucose)
Sucrose	92	61–67
Glucose	138	100
Fructose	32	23
Lactose	63	46
White bread	100	72
Pasta	46–88	33–64
Rice (polished)	68–104	49–75
Rice (parboiled)	58–78	42–56
Cornflakes	107–139	78–100
Potatoes	66–120	48–87
Beans	40–60	30–43
Lentils (red, dried)	25–45	18–33
Peas (green, frozen)	55–74	40–54
Oranges	44–73	32–53
Bananas	43–99	31–72
Apples	40–57	29–41

[a]GI has been calculated compared to white bread or glucose. The values are given as a mean or range. GI white bread/GI glucose = 1.38.
Source: Raben and Hermansen 2006.

rice are medium GI (56–69). High-GI foods (>70) include white bread, baked potatoes, and many breakfast cereals. The GI will be slightly higher when fruits are consumed as juice compared with whole fruit (Raben and Hermansen 2006). Because we consume a mixture of foods rather than a single food, **glycemic load** has been introduced to consider the quantity and quality of carbohydrates in a meal. The glycemic load (GL) is the glycemic index times the grams of carbohydrate in a serving of food.

Sugar Metabolism

Humans evolved to use glucose as their principal source of energy. Through the **glycolysis** pathway, glucose is converted under the "normal" condition of aerobic metabolism to two units of pyruvate, which is transported to the mitochondria where it enters the tricarboxylic acid (TCA) cycle. Metabolites can be completely oxidized to CO_2 and water with the release of energy via the TCA cycle. It is estimated that over 90% of food-derived energy is produced by TCA cycle oxidation (Gropper et al. 2009). Under anaerobic glycolysis, pyruvate will be reduced to lactate. Anaerobic glycolysis occurs in red blood cells because red blood cells do not contain mitochondria, where the TCA cycle is located within the mitochondrial matrix. Glycolysis takes place in virtually all tissues of the human body. The brain, the central nervous system, and red blood cells are particularly dependent on glucose as a nutrient. If blood glucose levels are low because of low carbohydrate consumption, glucose will be generated from noncarbohydrate substrates, such as lactate, pyruvate, and glycerol. This is known as **gluconeogenesis**. The metabolic fate of glucose is regulated through hormonal and enzyme activation/suppression, in accord with the body's energy needs. Normal blood glucose levels are maintained via the pancreatic hormones insulin and glucagon. Excess glucose can be stored in the body by conversion to glycogen (**glycogenesis**). The liver is the major site for glycogen synthesis and storage. The glycogen content of the liver is in the order of 7% (wet weight basis) and in muscle tissue 1%. Because of the larger mass of muscle tissue, approximately 75% of the body's glycogen is stored in skeletal muscle tissue. Through **glycogenolysis**, glucose will be cleaved from glycogen and subsequently used for energy via glycolysis. Glucose can also be the precursor for both the glycerol and fatty acid components of triacylglycerols, the metabolic process known as **lipogenesis**. Thus, excess carbohydrate consumption will lead to obesity.

Fructose is metabolized in the liver, where it is phosphorylated, and then enters the glycolytic pathway. Formation of fructose-1-phosphate is the major metabolic route to glycolysis. Formation of fructose-6-phosphate also occurs, but it is less significant because the reaction is slow and only occurs in the presence of high amounts of fructose. Galactose after being phosphorylated to galactose-1-phosphate in the liver can enter the glycolytic pathway, where it is converted to glucose-1-phosphate. The major dietary source of galactose is lactose. The hexose monophosphate shunt is another metabolic pathway that is available to glucose. In this pathway, pentose intermediates are formed that are critical for the synthesis of nucleic acids. Pentose sugars will be metabolized through this pathway. The sugar alcohol xylitol will be converted to xylulose in the liver and then metabolized via the pentose–phosphate shunt. Sorbitol is oxidized to fructose in the liver by sorbitol dehydrogenase and then metabolized as fructose.

The Large Intestine and the Digestive Process

The Colon

Carbohydrates that have not been hydrolyzed to monosaccharides and absorbed in the small intestine will empty into the cecum, the upper section of the colon. The colon has been referred to as the least understood organ in the human body. Following the cecum, the sequential regions or the colon are known as the ascending, transverse, descending, and sigmoid sections. The colon is approximately 5 feet long and larger in diameter than the small intestine. Passage of material through the colon takes from 12 to 70 hours, during which time materials are fermented and dehydrated. Approximately 90–95% of the water entering the colon is absorbed (Gropper et al. 2009). The colon also efficiently absorbs sodium, chloride, and potassium ions. Carbohydrate materials entering the colon include polysaccharides, oligosaccharides, and monosccharides that were not absorbed in the small intestine. These materials will be subject to fermentation by colonic bacteria, and fermentation products such as short-chain fatty acids will be absorbed.

Intestinal Microflora

Although there are some bacteria present in the mouth, stomach, and small intestine, it is the large intestine that contains an extremely large and diverse microbial population. An individual contains from 300 to 500 different species of bacteria, and in the order of 10^{11} or 10^{12} cells/g of luminal contents (Guarner and Malagelada 2003). Bacteria make up approximately 60% of dry mass of feces (Guarner and Malagelada 2003). The population includes bacteroides, bifidobacterium, eubacterium, clostridium, escherichia, enterobacter, lactobacillus, etc. Over 99% of the gut microflora are anaerobes. The microbial population is extremely variable from individual to individual, and subject to change within an individual by such factors as diet, medications, age, lifestyle, individual health, etc.

A symbiotic relationship exists between the host and the gut microflora, which perform several usual functions (Sears 2005). Energy is produced by fermentation of carbohydrates to short chain fatty acids (acetate, propionate, butyrate) that are used for energy and synthesis of important intermediates. The vitamins biotin and vitamin K are microbially produced and then absorbed across the colonic wall. The microflora help to prevent the growth of harmful pathogenic bacteria, and they help support a healthy immune system.

Probiotics are live microorganisms that are consumed as part of fermented foods, such as yogurt or dietary supplements; the intent is for their survival of the upper digestive tract and colonization of the gut with improved health of the host as a result. Lactic acid bacteria and bifidobacteria are the most common types of microbes used as probiotics. Possible health benefits include enhancing the body's immune function, managing lactose intolerance, treating antibiotic-associated diarrhea, lowering cholesterol, and preventing colon cancer and irritable bowel syndrome (Gropper et al. 2009). **Prebiotics** are nondigestible food ingredients that stimulate the growth and/or activity of colonic bacteria, which in turn benefits the health of the host (Roberfroid 2007). The pre-eminent prebiotic examples are fructooligosaccharides and inulin. Food sources for inulin are Jerusalem artichoke and chicory root. Although bananas and garlic contain some inulin, they are an insufficient dietary source. Inulin and fructoligosaccharides are isolated from plant sources or synthesized chemically for use as food ingredients. Other soluble dietary fiber, such as pectin, β-glucans, and polydextrose, may provide prebiotic effects.

A report of the **French Paradox** on the television program *60 Minutes* in 1991 stimulated research on the potential beneficial effects of red wine consumption. The term French Paradox is the observation that Frenchmen suffer a relatively low risk of coronary heart disease compared with American men, despite having a diet relatively rich in saturated fats. It was postulated that the protective effects could be attributed, in part, to phenolic secondary metabolites. Anthocyanin pigments have been shown to be very effective in scavenging free radicals in vitro. It was proposed that they prevented oxidation of cholesterol and other compounds in the bloodstream through free radical scavenging. However, <1% of

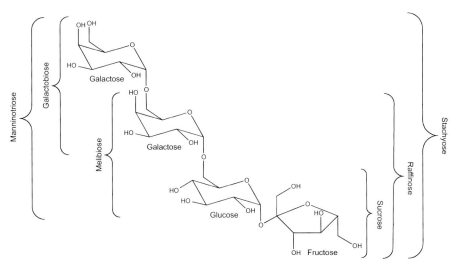

Figure 9.2   Structures of the raffinose/stachyose oligosaccharide family.

anthocyanins consumed are absorbed in the small intestine. There is evidence that colonic microflora degrade flavonoids to phenolic acids, which are absorbed and exert their protective effect through cell-signaling and gene regulation (Crozier et al. 2009).

Fate of Nonabsorbed Monosaccharides, Sugar Derivatives, and Oligosaccharides

Monosaccharides and sugar alcohols that are slowly absorbed may enter the colon when substantial quantities are consumed. They will be subject to fermentation, which can result in flatulence, and their hydrophilic nature can stimulate water absorption into the colon, resulting in diarrhea. The raffinose/stachyose family of oligosaccharides (Figure 9.2) are not hydrolyzed by the carbohydrase enzymes in the small intestines and not absorbed. They are found in legumes and are the cause of flatulence from bean consumption. Efforts have been made to reduce their concentration through plant-breeding and cooking/processing procedures such as soaking in cold water. Another therapy is oral consumption of an α-galactosidase-containing dietary supplement (Beano™).

Lactose and sucrose are the starting materials for food additives that deliver desired nutritional and functional properties. Lactulose (4-O-β-D-galactopyranosyl-β-D-fructose) is a laxative formed by alkaline isomerization of lactose (refer to Chapter 3).

Structure 9.3   (lactulose)

Lactitol (4-O-β-D-galactopyranosyl-β-D-glucitol) is produced by catalytic reduction of lactose.

Structure 9.4   (lactitol)

It is not hydrolyzed in the small intestine. It is a reduced-calorie bulking agent with a sweetness value approximately 60% of sucrose. Isomaltulose (6-O-α-D-glucopyranosyl-D-fructose, also known as Palitinose™) is produced by enzymatic isomerization of sucrose via microbial fermentation (Figure 9.3). It is hydrolyzed in the small intestine but absorbed at a slow rate. It has a sweetness level of 40–45% of sucrose and a low GI. Catalytic reduction of isomaltulose produces 6-O-α-D-glucopyranosyl-D-sorbitol and 1-O-α-D-glucopyranosyl-D-mannitol dihydrate (Figure 9.3). The mixture is known as isomalt and is an approved food additive in the United States, Europe, and many other countries. Although it is partially hydrolyzed by the carbohydrase enzymes in the small intestine, the rate is very low, resulting in approximately 10% hydrolysis (Sentko and Willibald-Ettle 2006). The remainder is subject to fermentation in the colon, and like most sugar alcohols, there is risk of gastric

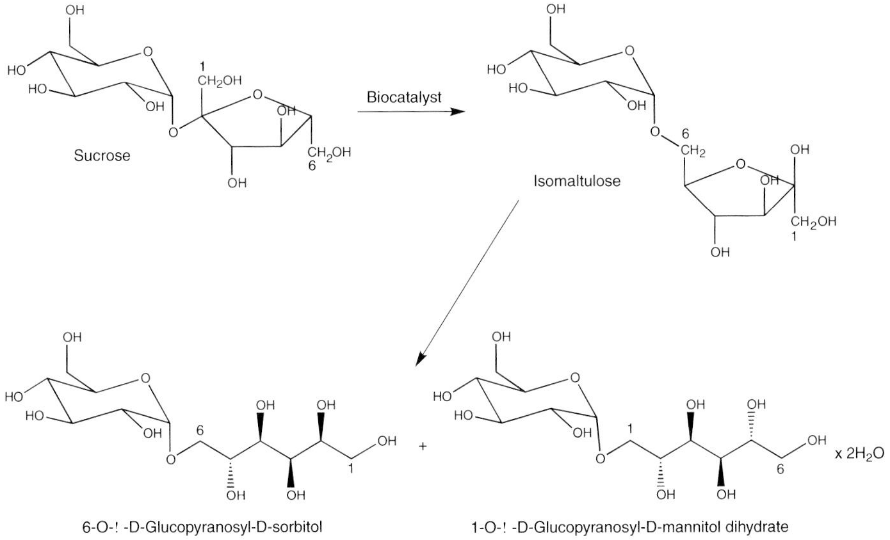

Figure 9.3 Formation of isomaltulose (Palitinose™) and 6-O-α-D-glucopyranosyl-D-sorbitol plus 1-O-α-D-glucopyranosyl-D-mannitol dihydrate (Isomalt) from sucrose.

distress. Its caloric value is approximately half that of glucose. Food applications include being an ingredient for hard candies, chocolates, chewing gum, fruit spreads, and breakfast cereals. The intense sweetener sucralose (1,6-dichloro-1-6-dideoxy-β-D-fructofuranosyl-4-chloro-4-deoxy–α-D-galactopyranoside; refer to Chapter 5) is not hydrolyzed in the small intestine.

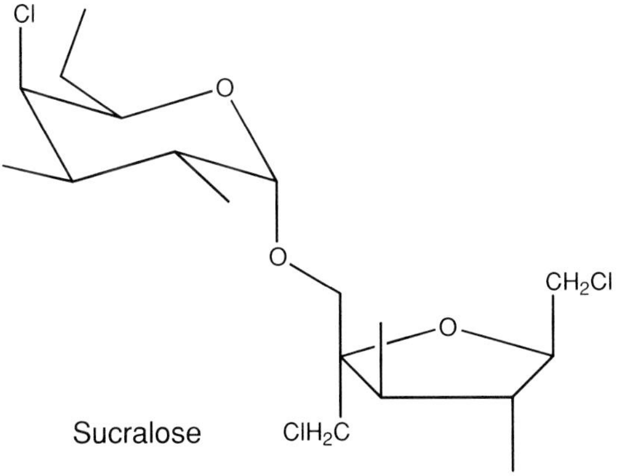

Sucralose

Structure 9.5   (sucralose)

Because of its high sweetness level (600× sucrose), the quantity of the additive consumed is very small; hence, risk of gastric distress is very low. Olestra, also known as Olean™, is a fat substitute produced by esterification of the free hydroxyl groups of sucrose with long-chain fatty acids. It was discovered and patented by Procter and Gamble.

Structure 9.6   (Olestra)

It is not absorbed and passes into the colon undigested without contributing calories or nutritive values to the diet. A major intended use was for high-fat foods such as potato chips. Procter and Gamble received approval from FDA for its use as a food additive in 1996 after a very lengthy petition process. A caveat that thrilled Procter and Gamble's marketing department was FDA's requirement of a warning label that Olestra "may cause abdominal cramping and loose stools." FDA removed the warning requirement in 2003; however, olestra was never widely accepted by consumers. Food use of olestra is banned in many countries.

Dietary Fiber

Dietary fiber consists of nondigestible carbohydrates and lignin that are found in plants. Thus, with the exception of those starch molecules that are hydrolyzed and absorbed in the small intestine, virtually all other polysaccharides are components of dietary fiber. Dietary fiber is further classified as to whether it is water-soluble or water-insoluble. Insoluble fiber includes cellulose, some hemicelluloses, and lignin. Soluble fiber includes pectin, gums, β-glucans, fructans, resistant starches, and some hemicelluloses. It is recognized that fiber is important for gastrointestinal tract function and for preventing and managing a variety of diseases. The United States National Academy of Sciences, Institute of Medicine, recommends that adults should consume 20–35 g of dietary fiber per day; however, the average American's daily intake is only 12–18 g (Higdon and Drake 2009). Nondigestible carbohydrates that have been extracted or isolated from natural sources are available as ingredients for foods and dietary supplements. These materials are classified as **functional fiber**. This category also includes synthetic materials such as polydextrose as well as some modified starches and cellulose derivatives. Chitin, which is an animal source, being isolated from the exoskeleton of crustacea, is included in this category.

The physiological effects of fiber in the gastrointestinal tract are as varied as the number of fiber components and determined largely by the type and amounts present (Gropper et al. 2009). Both solubility and water-holding capacity are important functional properties of dietary fiber. Water-soluble fibers tend to have a high water-holding capacity compared with cellulose and lignin. Materials with high water-holding capacity will increase fecal bulk and speed up fecal passage through the colon. Some water-soluble fibers will form viscous solutions that delay emptying of food from the stomach and provide a feeling of fullness (satiety), slow down the digestive process, and reduce the rate of nutrient absorption. Some fibers such as pectin, with its uronic acid functional groups, can complex with cations and influence mineral absorption. This can be positive or negative, with the more rapidly fermentable fibers having a favorable effect on mineral balance. The complexing of carotenoids and some phytochemicals with pectin and guar gum has been shown to substantially reduce the absorption of β-carotene, lycopene, lutein, and canthaxanthin (Gropper et al. 2009).

Carbohydrate Nutrition and Human Health

Dietary carbohydrates have several roles that are important in achieving and maintaining optimal health. The cause of dental caries has long been associated with consumption of high amounts of sugars. Bacteria (*Streptococcus mutans* and *Lactobacillus*) ferment sucrose, fructose, and glucose to produce lactic and other acids, which damage the tooth enamel. Good dental hygiene and use of fluoride-containing toothpaste and drinking water are effective preventative measures. Individuals may suffer from metabolic disorders such as **lactose intolerance**, a metabolic disorder with which individuals do not synthesize sufficient lactase to hydrolyze ingested lactose. Lactase activity is high in infants with activity decreasing a few years after weaning. Lactose intolerance is particularly prevalent among African Americans, Jews, Arabs, Greeks, and some Asians (Gropper et al. 2009). Alternative plant-based milk products that are lactose-free are available, and lactase supplements are also available.

A very serious metabolic disorder is **diabetes mellitus type 1** (juvenile diabetes), in which autoimmune destruction of insulin-producing beta cells of the pancreas occurs. The conditions leads to elevated blood and urine glucose levels and is fatal unless treated with insulin. In **diabetes mellitus type 2** (adult-onset diabetes), elevated blood glucose occurs as a result of insulin resistance and relative insulin deficiency (Robbins et al. 2005). It contrasts with type 1 diabetes in that the problem is with the cells' response to insulin, rather than insulin production. The incidence of type 2 diabetes mellitus is increasing at an alarming rate worldwide (Raben and Hermansen 2006), largely because of lifestyle and the epidemic of obesity. Being overweight or obese is considered the major cause of type 2 diabetes in genetically predisposed individuals (Raben and Hermansen 2006). With its initial onset, it can often be managed through exercise and dietary modification that incorporates high fiber and foods with low GI.

Obesity has become a major health crisis in the United States and much of the Western world. The cause of this epidemic is attributed to a lifestyle characterized by lack of exercise and a high-fat diet that has low amounts of fruits and vegetables. Consumption of high amounts of simple sugars, particularly in the form of sweetened beverages, has been implicated as a contributing factor. Several studies (Ludwig 2003) have shown that consumption of low-GI foods delayed the return of hunger, decreased subsequent food intake, and increased satiety when compared with high-GI foods. Coronary heart disease is closely linked with obesity, and type-2 diabetes also increases the risk of heart disease. The American Heart Association recommends reduction in the consumption of sugars added to foods during processing, preparation, or at the table to no more than 100 calories per day for women and 150 for men (Johnson et al. 2009). High consumption of dietary fiber is associated with significant reduction in coronary heart disease (Higdon and Drake 2009). One of the cardioprotective effects of dietary fiber is its ability to lower cholesterol levels; however, the effect of fiber consumption on blood glucose and insulin responses may also contribute. In regard to cancer, good carbohydrate nutrition may be helpful, but as a preventative measure it is probably secondary. In humans, studies have in general not shown an association between high consumption of added sugars and cancer (Raben and Hermansen 2006). A number of studies have shown that the incidence of colorectal cancer was lower in people with high fiber intakes (Trock et al. 1990; Howe et al. 1992). Several clinical trials, however, have failed to demonstrate a protective effect of fiber consumption on the reoccurrence of precancerous polyps (Higdon and Drake 2009). Although some studies have shown an inverse relationship between dietary fiber intake and the incidence of breast cancer, there is not sufficient evidence at present that high fiber intakes significantly decrease the risk of breast cancer in women.

There is a ready audience for books on diet programs designed for weight loss that will result in improved health and physical appearance. The *Atkins Diet Revolution* became a best seller in 1972 (Atkins 1972). The low-carbohydrate diet operates on the premise that deriving calories from fat rather than carbohydrate will "burn" more calories through ketosis. It is also proposed that the onset of hunger will be delayed because of the longer duration for digestion of fats and proteins. The popularity of the diet surged again in 2003 and 2004. The *Pritikin Program for Diet and Exercise* (Pritikin and McGrady 1979) includes a daily aerobic exercise program combined with a diet emphasizing unprocessed or minimally processed fruits, vegetables, legumes, whole cereal grains, lean meat, and seafood. *The South Beach Diet* (Agaston 2003) emphasizes consumption of low-GI foods along with foods rich in unsaturated fatty acids and omega-3 fatty acids.

Amateur and professional athletes are eager to follow the advice of sports nutritionists in order to optimize their competitive performance. For endurance events, marathon runners and cross-country skiers often follow the regimen for **carbohydrate loading** (or carb-loading). The intent of this process is to elevate glycogen levels in skeletal muscle (Gropper et al. 2009). There are variations regarding the prescribed time table, but basically a period of intense exercise is done approximately a week before the competitive event to deplete muscle glycogen. This is followed by a tapering down of exercise combined with a short period (2–3 days) of low carbohydrate (10–50%) consumption; this is followed by 3 days of high carbohydrate diet (70–90%). Glycogen levels can be elevated 20–40% above normal. The timing of pre-event meals is critical. Carbohydrate consumption too close to the event can result in insulin release and rapid reduction of plasma glucose. Having the final meal 3–4 hours before the event empties the stomach. For endurance events, the meal should be high in complex carbohydrates and low in fat. Consumption of an isotonic beverage containing glucose 15–20 minutes before the event provides extra dietary glucose. Consuming glucose-containing beverages during a prolonged event helps to maintain fluid balance and glucose levels. The isotonic beverage should be cool, not cold, and contain glucose or polyglucose. Substantial amounts of fructose should not be included because of its slow absorption rate.

Vocabulary

Active transport—Process where sugars are absorbed against a concentration gradient; galactose and glucose are absorbed by active transport.

Carbohydrate loading—Regimen of exercise and diet designed for endurance athletes with the purpose of elevating the levels of muscle glycogen

Diabetes mellitus type 1 (juvenile diabetes)—Metabolic disorder where autoimmune destruction of insulin-producing beta cells of the pancreas occurs; also referred to as IDDM (insulin-dependent diabetes mellitus)

Diabetes mellitus type 2 (adult-onset diabetes)—Metabolic disorder where elevated blood glucose occurs as a result of insulin resistance and relative insulin deficiency; also referred to as NIDDM (non-insulin-dependent diabetes mellitus)

Dietary fiber—Nondigestible carbohydrates and lignin that are found in plants

Facilitated transport—Mechanism for fructose absorption, which occurs down a concentration gradient, independent of glucose concentration, and is facilitated by specific transporters

French Paradox—Term describing the observation that Frenchmen suffer a relatively low risk of coronary heart disease compared with American men, despite having a diet relatively rich in saturated fats

Functional fiber—Nondigestible carbohydrates that have been isolated, extracted, or manufactured and have been shown to have beneficial physiological effects in humans

Gluconeogenesis—Metabolic pathway that results in the generation of glucose from noncarbohydrate carbon substrates such as lactate, pyruvate and glycerol

Glycemic index (GI)—A measure of the increase in blood glucose level over the base-line level during a 2-hour period following consumption of 50 g of available carbohydrate

Glycemic load (GL)—The glycemic index times the grams of carbohydrate in a serving of food

Glycogenesis—Glycogen synthesis

Glycogenolysis—Conversion of glycogen polymers to glucose monomers

Glycolysis—Metabolic pathway that converts glucose to pyruvate with the release of energy

Lactose intolerance—A metabolic disorder where individuals do not synthesize sufficient lactase to hydrolyze ingested lactose

Lipogenesis—Metabolic process where glucose is converted to triacyglycerols

Passive absorption—Process where sugar absorption takes place according to differences in osmotic pressure; mannose, pentose sugars, L-sugars, and sugar alcohols are absorbed by passive diffusion.

Prebiotics—Nondigestible food ingredients that stimulate the growth and/or activity of colonic bacteria, which in turn benefits the health of the host

Probiotics—Live microorganisms that when administered in adequate amounts confer a health benefit to the host

References

American Diabetes Association. 2010. *The glycemic index of foods.* Available from: http://www.diabetes.org. Accessed June 28, 2010.

Agaston A. 2003. *The South Beach diet: the delicious, doctor-designed, foolproof plan for fast and healthy weight loss.* New York: St Martin's Press.

Atkins RC. 1972. *Dr. Atkins' diet revolution: the high calorie way to stay thin forever.* New York: D. McKay.

BeMiller JM, Huber KC. 2008. Carbohydrates. In: Damodaran S, Parkin KL, Fennema OR, editors. *Fennema's food chemistry, 4th ed.*, Boca Raton, FL: CRC Press, Taylor & Francis, pp. 83–154.

Crozier A, Jaganath IB, Clifford MN. 2009. Dietary phenolics: chemistry, bioavailability and effects on health. *Nat Prod Rep* 26:1001–1043.

Foster-Powell K, Holt SH, Brand-Miller JC. 2002. International table of glycemic index and glycemic load. *Am J Clin Nutr* 76:5–56.

Gropper SS, Smith JL, Groff JL. 2009. *Advanced nutrition and human metabolism.* Belmont, CA: Wadsworth Cengage Learning.

An excellent reference that is widely used as a recommended text in nutrition courses.

Guarner F, Malagelada J-R. 2003. Gut flora in health and disease. *Lancet* 360:512–519.

Higdon J, Drake VJ. 2009. Micronutrient Information Center. Available at: http://lpi.oregonstate.edu/infocenter/phytochemicals/fiber.

The Micronutrient Information Center located on the Linus Pauling Institute website is an excellent resource for current peer-reviewed information on vitamins and micronutrients.

Howe GR, Benito E, Castelleto R, Cornee J, Esteve J, Gallagher RP, Iscovich JM, Deng-ao J, Kaaks R, Kune GA. 1992. Dietary intake of fiber and decreased risk of cancers of the colon and rectum: evidence from the combined analysis of 113 case-control studies. *J Natl Cancer Inst* 84:1887–1896.

Jenkins DJA, Wolever TMS, Taylor RH, Barker H, Fielden H, Baldwin JM, Bowling AC, Newman HC, Jenkins AL, Goff DV. 1981. Glycemic index of foods: a physiological basis for carbohydrate exchange. *Am J Clin Nutr* 34:362–366.

Johnson RK, Appel LJ, Brands M, Howare BV, Lefevre M, Lustig RH, Sacks F, Steffen LM, Wylie-Rosett J. 2009. Dietary sugars intake and cardiovascular health: a scientific statement from the American Heart Association. *Circulation* 120:1011–1020.

Khanna S, Parrett A, Edwards CA. 2006. Nondigestible carbohydrates: nutritional aspects. In: Eliasson A-C, editor. *Carbohydrates in food, 2nd ed.*, Boca Raton, FL: CRC Press, Taylor & Francis, pp. 273–303.

Ludwig DS 2003. Dietary glycemic index and the regulation of body weight. *Lipids* 38:117–121.

Pritikin N, McGrady PM Jr. 1979. *The Pritikin program for diet and exercise.* New York: Grosset and Dunlap.

Raben A, Hermansen K. 2006. Health aspects of mono- and disaccharides. In: Eliasson A-C, editor. *Carbohydrates in food, 2nd ed.*, Boca Raton, FL: CRC Press, Taylor & Francis, pp. 89–127.

Robbins SL, Cotran RS, Kumar V, Abbas AK, Fausto N. 2005. *Robbins and Cotran pathologic basis of disease, 7th ed.*, Philadelphia: Elsevier.

Roberfroid M. 2007. Prebiotics: the concept revisited. *J Nutr* 137:830S–837S.

Sears CL. 2005. A dynamic partnership: celebrating our gut flora. *Anaerobe* 11:2247–2251.

Sentko A, Willibald-Ettle I. 2006. Isomalt. In: Mitchell H, editor. *Sweeteners and sugar alternatives in food technology.* Ames, IA: Blackwell Publishing, pp. 177–204.

Trock B, Lanza E, Greenwald P. 1990. Dietary fiber, vegetables, and colon cancer: critical review and meta-analyses of he epidemiologic evidence. *J Natl Cancer Inst* 82:650–661.

Whistler RL, BeMiller JM. 1997. *Carbohydrate chemistry for food scientists.* St. Paul, MN: Eagan Press.

Zapsalis C, Beck RA. 1985. Food chemistry and nutritional biochemistry. New York: John Wiley & Sons.

Appendices

Unit 1 Laboratory/Homework Exercise—Building Molecular Models of Sugar Molecules

Expected Outcomes

- To build molecular models of sugars using commercial molecular model kits
- Be able demonstrate, using molecular models, the meaning of conformers, epimers, anomers, enantiomers, tautomers, axial and equatorial positions, steric restrictions, and ring strain

Materials

Molecular model kit. We use Darling MOLECULAR VISIONS™ Organic, Inorganic, Organometallic, ISBN 978-09648837-1-O. Available from Darling Models, Inc. for $22. http://www.darlingmodels.com

Assignment

Using a molecular model kit, carry out the following operations:

- Build a model of β-D-glucopyranose.

Structure: β-D-glucopyranose

Food Carbohydrate Chemistry, First Edition. Ronald E. Wrolstad.
© 2012 John Wiley & Sons, Inc. Published 2012 by John Wiley & Sons, Inc.

- Refer to Figure 1.6. Place β-D-glucopyranose into 8 strainless conformers. Identify the C-1 and 1-C conformers.

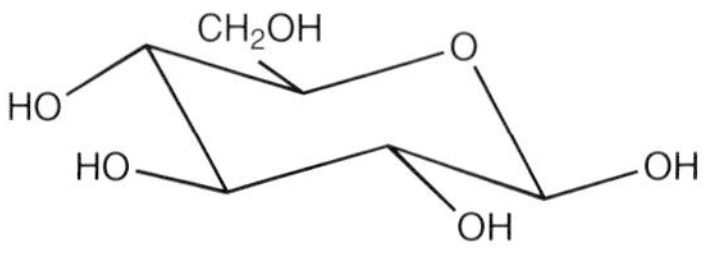

C1

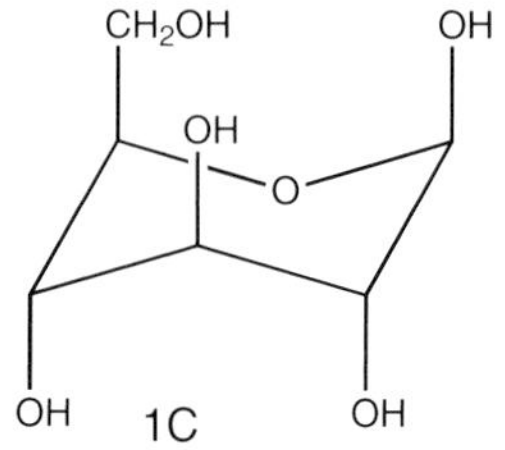

1C

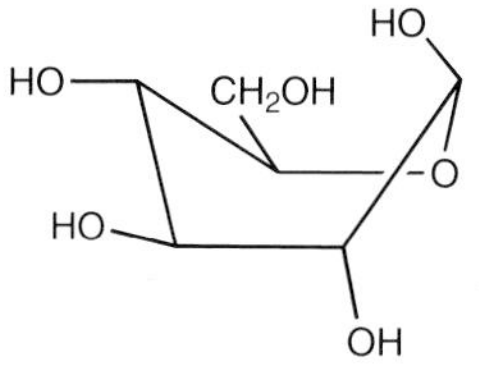

B1

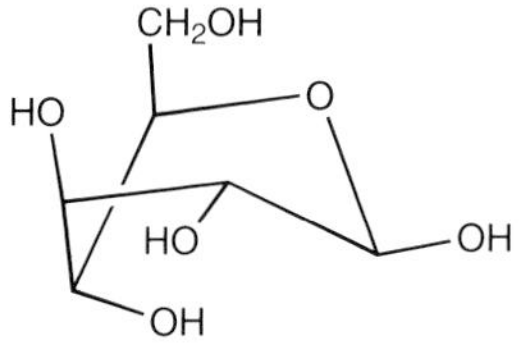

1B

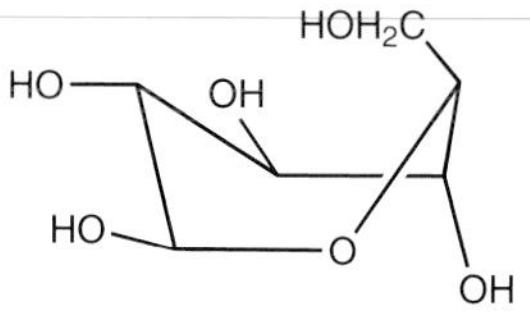

B2

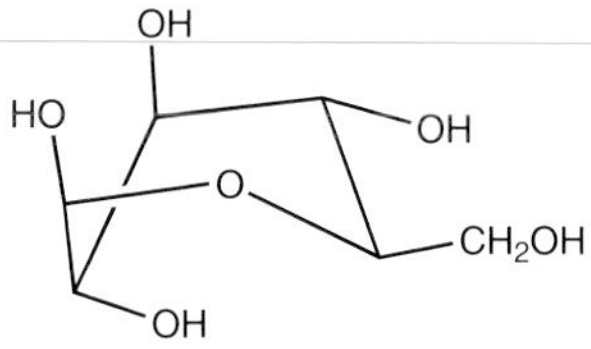

2B

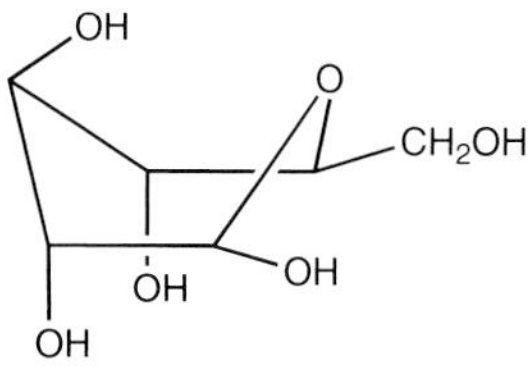

B3

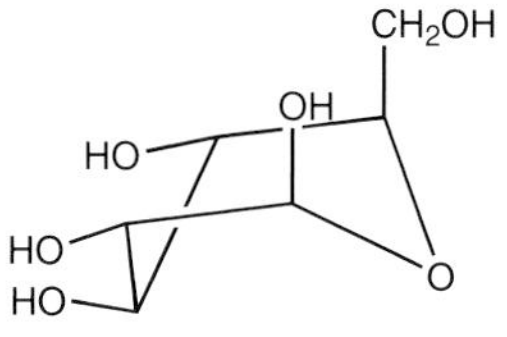

3B

- Identify the anomeric carbon atom. Make the α-anomer.
- Make the 4-epimer of glucose.
- Make the enantiomer of β-D-glucopyranose.
- Make β-D-glucofuranose.
- Make the acyclic form of D-glucose.
- Build L-ascorbic acid. Why does ascorbic acid prefer the furanose to the pyranose form?

Structure: Ascorbic acid

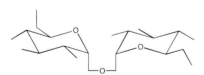

• Refer to Figure 1.9 and construct the following disaccharides: sucrose, α- and β-lactose, α- and β-maltose, α- and β-cellobiose, and trehalose.

Sucrose

α-Lactose

β-Cellobiose

β-Maltose

Trehalose

Unit 2 Homework Exercise—Recognizing Hemiacetal, Hemiketal, Acetal, and Ketal Functional Groups

Identifying hemiacetal, hemiketal, acetal, and ketal functional groups is critical to determining whether a sugar is reducing or nonreducing. The first step is to locate the anomeric carbon atom, which will be adjacent to the ring oxygen. Indicate the location of all hemiacetal, hemiketal, acetal, and ketal functional groups in the compounds below:

β-D-fructopyranose structure:

Methyl-β-D-glucopyranoside structure:

Food Carbohydrate Chemistry, First Edition. Ronald E. Wrolstad.
© 2012 John Wiley & Sons, Inc. Published 2012 by John Wiley & Sons, Inc.

β-maltose structure:

Sucrose structure:

Answers to Exercise

Hemiketal

Acetal

Acetal

Hemiacetal

Acetal

Ketal

Unit 3 Laboratory/Homework Exercise—Specification of Conformation (C-1 or 1-C), Chiral Family (D or L), and Anomeric Form (α or β) of Sugar Pyranoid Ring Structures

Supplementary materials

Molecular models of β-D-glucopyranose, β-L-glucopyranose, and the D- and L-isomers of 1,1-dimethyltetrahydropyran.

Prologue

While on sabbatical leave at Cornell University in 1980, I had the privilege of auditing Bob Shallenberger's course in Food Carbohydrate Chemistry. Following his lecture on chirality and optical activity of pyranoid sugars, we discussed the difficulties of correctly drawing L-forms and various sugars with nonconventional orientation on the printed page. Bob suggested that it should be possible to check whether assignments of D and L, and α and β were correct by an algebraic procedure. In the next 2 weeks, Bob, Laurie Kerschner, his graduate student, and myself set about developing and testing a system for determining whether a sugar pyranose structure was C-1 or 1-C, D or L, and α or β. A manuscript

Food Carbohydrate Chemistry, First Edition. Ronald E. Wrolstad.
© 2012 John Wiley & Sons, Inc. Published 2012 by John Wiley & Sons, Inc.

was submitted and published shortly thereafter (Shallenberger et al. 1981).

An important concept is that chirality (or optical activity) depends on whether the molecule as a whole is asymmetric. Xylitol contains asymmetric carbon atoms, but it is not chiral because the molecule as a whole is symmetrical. 1-1-Dimethyltetrahydropyran contains no asymmetric carbon atoms but is optically active. The molecule exists in chair forms, and the two mirror images are not superimposable, that is, they cannot be stacked on top of each other. (If the molecules were planar, they would be superimposable.)

Structure 1,1-dimethyltetrahydropyran:

Determination of Chair Conformation

- Locate the anomeric carbon atom and determine if the numbering sequence is clockwise or counterclockwise. If clockwise, operand **n** is $+$; if counterclockwise, operand **n** is $-$.
- Observe whether the "puckered" ring oxygen atom lies above the plane of the ring or below. If above, operand **p** is $+$; if below, **p** is $-$.
- Multiply **n** by **p**. If the product is positive, the conformation is C-1; if negative, it is 1-C.

Note: The interrelationship between clockwise/counterclockwise and orientation of the ring oxygen can be demonstrated with a model of β-D-glucopyranose in the C-1 conformation. Holding the model in space with the conventional orientation of the ring oxygen being above the plane of the ring, the carbons will have a clockwise orientation. Turning the molecule 180° without twisting any bonds, the oxygen will be below the plane of the ring and the numbering will be counterclockwise. The shape of the model has not been changed in any way. It simply has a different orientation in space. It will still be β-D-glucopyranose in the C-1 conformation.

Determination of Chiral Family

- Locate the reference carbon atom contained within the ring and determine whether the bulky substituent (either OH or CH₂OH) is equatorial or axial. If the susbstituent has an equatorial disposition, operand **r** is +; if axial, operand **r** is −.
- Multiply (**np**) by **r**. When the product is positive, the chiral family is D; when it is negative, the chiral family is L.

Determination of Anomeric Form

- Determine whether the hydroxyl or OR substituent on the anomeric carbon is equatorial or axial. If it is equatorial, operand **a** is +; if it is axial, operand **a** is −.
- Multiply (**np**) by (**npr**) by (**a**). When the product is +, the anomer is β; when the product is −, the anomer is α.

*Note: Algebraic short cut—Since n^2 and p^2 will always be positive, simply multiply (**ra**). As above, + = β and − = α.*

Exercise: Calculating and Specifying Multiple Chirality of Sugar Pyranoid Ring Structures

Structure β-D-glucopyranose-C1:

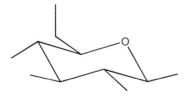

Operands			
n	p	r	a
+	+	+	+

(np) = (+), therefore C-1
(np)(r) = +, therefore D
(np)(npr)(a) = +, therefore β

Structure β-D-fructopyranose-1C:

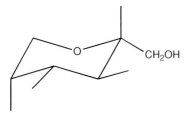

n	p	r	a

(np) =
(np)(r) =
(np)(npr)(a) =

Structure β-D-levoidosan-1C:

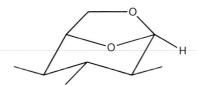

n	p	r	a

(np) =
(np)(r) =
(np)(npr)(a) =

Structure α-L-xylopyranose-1C:

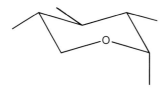

n	p	r	a

(np) =
(np)(r) =
(np)(npr)(a) =

Structure β-L-glucopyranose-1C:

n	p	r	a

(np) =
(np)(r) =
(np)(npr)(a) =

Answers to Exercise

Structure β-D-fructopyranose-1C:

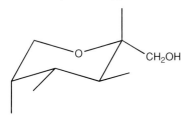

n p r a = + − − −, therefore β-D-1C

Structure β-D-levoidosan-1C:

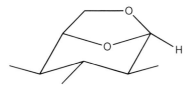

n p r a = + − − −, therefore β-D-1C

Structure α-L-xylopyranose-1C:

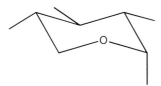

n p r a = − + + −, therefore α-L-1C

Structure β-L-glucopyranose-1C:

n p r a = − + + +, therefore β-L-1C

References

Shallenberger RS. 1982. *Advanced sugar chemistry*. Westport, CT: AVI Publishing.

Shallenberger RS, Wrolstad RE, Kerschner LE. 1981. Calculation and specification of the multiple chirality displayed by sugar pyranoid ring structures. *J Chem Educ* 58: 599–601.

Unit 4 Demonstration of the Existence of Plane-Polarized Light and the Ability of Sugar Solutions to Rotate Plane-Polarized Light

Materials

- Light source that will project a beam of visible light. We use a microscope lamp source, but a hand-held flashlight will suffice.
- Two polarizing filters. Sheets of polarizing film (12 × 21 cm) were affixed with tape to 15 × 24-cm hardboard frames that had a 10 × 16.5-cm window. It is critical that the polarizing film be mounted with the grid oriented in the same direction. Visible linear polarizing film is available from Edmund Optics (http://www.edmundoptics.com).
- Two transparent containers to serve as cells. One contains distilled water and the other corn syrup. Plastic utility boxes (500 mL capacity, 191 × 15.6 × 3.7 cm) used for electrophoresis will serve the purpose. Cells available from VWR International (https: www.vwr.com).

Demonstration

The demonstration is most effective in dim light; overhead lights should be turned off. Direct a beam of light toward the audience. Place filter in front of light beam. Light emitted will be traveling in a single plane. Place the second filter so that it is oriented the same as the first filter. Plane-polarized light will be emitted through the second filter. Slowly rotate the second filter 180°. Very little, if any, light will be emitted from the second

---

Food Carbohydrate Chemistry, First Edition. Ronald E. Wrolstad.
© 2012 John Wiley & Sons, Inc. Published 2012 by John Wiley & Sons, Inc.

filter because the second filter has effectively blocked the light traveling in a single plane.

Place the cell containing distilled water between the two filters. Rotate the second filter. The effect will be the same as observed with air.

Place the corn syrup–containing cell between the two filters. Rotate the second filter. The angle with minimal light emission will not be 180°. The corn syrup solution has rotated plane-polarized light. Different colors will be observed. Light will appear to be refracted because corn syrup will rotate light of different wavelengths to varying degrees.

If a filter that emitted light at 589 nm was placed between the light beam and the first filter, the system would be a model of a polarimeter. The first filter serves as a polarizer, and the second filter as the analyzer.

Unit 5 Laboratory Exercise—Sugar Polarimetry

Expected Outcomes

- To become proficient in using a polarimeter
- To be able to calculate $[\alpha]_D^T$ and understand the meanings of d, l, D, L, dextrorotary, levorotary, specific rotation, optical activity, mutarotation, and sucrose inversion

Materials

- Polarimeter and 1-dm polarimeter cells
- Six 10% (w/v) unknown sugar solutions in coded 100-mL screw-cap bottles. The coded samples prepared 24 hours ahead include sucrose, fructose, glucose, invert sugar, and xylitol. The sixth unknown is glucose, freshly prepared just prior to the laboratory session.
- Two liters 0.05 M acetate buffer, pH 4.7 [255 mL 0.1 M acetic acid (6.0 g/L) + 245 mL 0.1 M NaAc (8.2 g $NaC_2H_3O_2$ or 13.6 g $NaC_2H_3O_2.3H_2O$)] diluted to 1 L.
- One liter 10% sucrose solution (0.29 M) in acetate buffer (molecular weight sucrose = 342)
- Commercial food-grade liquid invertase

Note: Invertase concentration for experiment should be pretested. We recommend a high concentration to accelerate the reaction so experiment can be done in less than 1 hour.

Food Carbohydrate Chemistry, First Edition. Ronald E. Wrolstad.
© 2012 John Wiley & Sons, Inc. Published 2012 by John Wiley & Sons, Inc.

Experimental

Part 1—Sugar Polarimetry

Become familiar with operating the polarimeter by observing the field without any sample in the sample port. Half of the field will be "bright" and the other half "dim." Adjust the polarimeter so the right and left halves of the field are dim and indistinguishable.

Determine α for water and the coded unknown sugar solutions. Calculate $[\alpha]_D^T$ and identify the coded samples from the values given in Supplemental Table 5.1. *Note:* The freshly prepared glucose solution will be undergoing mutarotation; hence, its α value will be changing throughout the laboratory period. For correct identification, values may need to be checked at subsequent times. Identify whether the freshly prepared glucose sample was made from crystalline α-D-glucose or β-D-glucose.

$$[\alpha]_D^T = \alpha/c \times l \tag{5.1}$$

α = Observed rotation; c = concentration, g/mL; l = pathlength in dm; D = D line of Na (589.3 nm); T = temperature

Appendix Table 5.1   Specific Rotation of Sugars[a] (Anhydrous Basis)

Sugar	"Initial" $[\alpha]_D^T$	Final $[\alpha]_D^T$
α-D-Glucose	+112°	+52°
β-D-Glucose	+19°	+52°
α-D-Galactose	+144°	+80°
β-D-Galactose	+52°	+80°
α-D-Mannose	+34°	+15°
β-D-Mannose	−17°	+15°
β-D-Fructopyranose	−133°	−92°
β-D-Fructofuranose	−21° (−17°)	−92°
α-D-Xylose	+92°	+19°
β-D-Xylose	−20°	+19°
α-D-Lactose	+90°	+55°
β-D-Lactose	+35°	+55°
β-D-Maltose	+118°	+136°
α,α-Trehalose	+178°	+178°
Sucrose	+66.5°	+66.5°
Raffinose	+124°	+124°
Stachyose	+148°	+148°
Invert Sugar		−20°

Modified from Shallenberger and Birch 1975.

Part 2—Sucrose Inversion

Thoroughly mix 1 mL of enzyme solution with 100 mL 0.29 M sucrose solution, pH 4.7. Place solution in polarimeter cell and read every 5 min. Plot $[\alpha]_D^T$ vs. time. Calculate percent inversion after 5, 10, and 30 min.

$$\% \text{ Inversion} = [\alpha]_D^T @ \text{ time } 0 - [\alpha]_D^T @ \text{ time } t/86.5 \times 100$$
$$(\text{Maximum change in } [\alpha]_D^T = +66.58 - (-20°) = 86.5°) \tag{5.2}$$

Determine initial reaction velocity (mMol sucrose hydrolyzed/min/L).

Suggested approach:

- Plot $[\alpha]_D^T$ vs. time.
- Draw straight line tangential to curve.
- Calculate %inversion at T_1 and T_2.

$$\text{Reaction Velocity} = \%\text{Inversion} @ T_2 - \%\text{Inversion} @ T_1/100$$
$$\times 290 \, \text{mM/L}/T_2 - T_2 \tag{5.3}$$

References

Shallenberger RS, Birch GG. 1975. *Sugar chemistry*. Westport, CT: AVI Publishing.

Unit 6 Laboratory Exercise or Lecture Demonstration— The Fehling's Test for Reducing Sugars

Expected Outcomes

- To determine whether a sugar is reducing or nonreducing by the Fehling's test
- To understand the concept of dextrose equivalency as applied to maltodextrins

Materials

Fehlings reagent

- Solution A: 34.6 g crystalline $CuSO_4$ in 500 mL H_2O
- Solution B: 173 g NaK tartrate[a] and 50g NaOH in 500 mL H_2O
- Sucrose, glucose, fructose, lactose, xylitol, starch
- Series of selected maltodextrins to be tested as unknowns; suggested dextrose equivalency (DE) of samples: 05, 10, 20, and 36

Note: It has been our experience that the technical applications departments of commercial starch companies will supply complementary samples of starches, modified starches, and maltodextrins for educational purposes.

- Test tubes, test tube holder, 250-mL beaker, hot plate, boiling chip, 5-mL pipette, pipette pump

---

Food Carbohydrate Chemistry, First Edition. Ronald E. Wrolstad.
© 2012 John Wiley & Sons, Inc. Published 2012 by John Wiley & Sons, Inc.

Background Information

The classical Fehling's test for reducing sugars is to first mix equal portions of solutions A and B (e.g., prepare a stock solution by combining 25 mL A and 25 mL B). Add 100 mg of substance to be tested to test tube containing 5 mL of Fehling's solution. Heat in boiling water bath for 5–10 min. A positive reaction is the formation of green, yellowish-orange, or red precipitate.

Maltodextrins are intermediate starch hydrolysate products having varying functional properties (sweetness, solubility, viscosity, etc.) dependent on the molecular size. Their reducing power is an index of their molecular weight and is expressed as dextrose equivalency (DE), the percentage of reducing sugars in s starch hydrolysate on a dry weight basis (e.g., glucose = 100 and starch = 0).

Experimental

Add 5 mL Fehling's stock solution to labeled test tubes, then add approximately 100 mg of substance to be tested to respective tubes. (Quantity of sugar does not need to be precisely weighed; it can be estimated.) Place samples in boiling water bath, and record the time for reaction to occur.

Apply the Fehling's test to the four unknown maltodextrin samples and identify the DE of each unknown.

[a]The purpose of the NaK tartrate is to buffer the solution in the alkaline range to prevent hydrolysis of nonreducing sugars. [Sodium bitartrate (cream of tartar) is acid, pH 3.9.]

References

Joslyn MA. 1970. *Methods in food analysis, 2nd ed*. New York: Academic Press.

Unit 7 Laboratory Exercise— Student-Designed Maillard Browning Experiments

Expected Outcomes

- To design and conduct an appropriate experiment that tests a hypothesis related to Maillard browning
- To draw appropriate conclusions from the experimental data

Prologue

This exercise is planned to give students, working in teams of two to three individuals, experience in designing an experiment that tests various hypotheses relating to Maillard browning. Teams selected a hypothesis and submitted a written experimental plan to the instructor for approval one week prior to conducting the experiment. This allows for experiment modification to ensure that it includes appropriate controls, and for it to realistically be completed in a laboratory session of 3 hours. It also helps in planning for sufficient laboratory materials.

Students were given the following advice:

- Keep the experimental plan simple.
- Consider the experiment to be a "preliminary" experiment, that is, an initial experiment to test conditions rather than a complete experiment suitable for publication.
- To minimize the number of samples, replication is not required.
- Maillard browning can be accelerated by conducting the experiment at 100°C. In previous years, students had successful results by reacting 1.0 M sugar solution with 1 mL of 1 M glycine solution at 100°C.

Food Carbohydrate Chemistry, First Edition. Ronald E. Wrolstad.
© 2012 John Wiley & Sons, Inc. Published 2012 by John Wiley & Sons, Inc.

Appendix Table 7.1 A_w of Glycerol/H_2O Solutions[a]

Glycerol (g)	H_2O (g)	Aw
0	100	1.0
20	80	0.90
32.3	67.7	0.86
70	30	0.63
85	15	0.37

Source: [a]Garzon and Wrolstad 2001.

- Use absorbance at 420 nm as a measurement of browning.
- Assume that Maillard browning will stop at $0°C$. (Test tubes containing the reaction solution can be withdrawn from a boiling water bath and plunged in an ice bath.)
- Measurement of the rate of browning will provide more reliable information than measurement at a single end point.
- Sorbitol can serve as a replacement for reducing sugars in controls.
- The rate of reaction can be accelerated by testing a pentose sugar rather than a hexose sugar.

Materials

- 1.0 M solutions of glucose, xylose, ribose, maltose, fructose, sucrose, and sorbitol (MW of hexose sugars = 180, pentose sugars = 150, maltose and sucrose = 342, glycine = 75).
- 1.0 M glycine solution
- 1 N NaOH and 1 N HCl for pH adjustment
- Ascorbic acid
- Glycerol

Equipment

- Visible spectrophotometer (Spectronic 20) and cells
- Hot plate
- 30-mL screw-cap test tubes and test tube rack
- Hot plate, 1 L beaker, and boiling chip
- Ice bath
- pH meter

Appendix Table 7.2   Composition of Kiwifruit Juice[a]

Reducing sugars	8.7 g/100 mL
Sucrose	0.92 g/100 mL
Ascorbic acid	98 mg/100 mL
Major amino acids[b]	213 mg/100 mL
pH	3.5

Source: [a]Wong and Stanton 1989.
[b]L-Glutamic acid, L-aspartic acid, L-glutamine, L-asparagine,
L-arginine.

Hypotheses

- The rate of Maillard browning is lowest at pH 3.0.
- Sucrose will undergo Maillard browning at pH 2.0, but will not at pH 8.0.
- The rate of Maillard browning increases with reduction of A_w.
- Equal molar concentrations of a reducing disaccharide will brown at a slower rate than a monosaccharide.
- Ascorbic acid contributes more to nonenzymatic browning in kiwifruit juice than reducing sugars.

References

Garzon GA, Wrolstad RE. 2001. The stability of pelargonidin-based antho-cyanins at varying water activity. *Food Chem* 75: 185–196.
Wong M, Stanton DW. 1989. Nonenzymic browning in kiwifruit juice concen-trate systems during storage. *J Food Sci* 54: 669–673.

Unit 8 Laboratory Exercise or Lecture Demonstration— Microscopic Examination of Starch

Expected Outcomes

- To prepare microscope slides of starch granules and examine their appearance under ordinary and polarized light
- To observe the staining of starch with iodine and KOH-assisted starch gelatinization with a light microscope
- To understand the meaning of "birefringence" and "cross-striations" and their relationship to crystallinity and gelatinization

Materials and Equipment

- Light microscope equipped with polarizing filters
- Microscope slides, cover slips, filter paper
- 1% aqueous suspensions of various starches (e.g., potato, corn, waxy maize, wheat, starch, tapioca, pregelatinized and modified corn starches)
- 0.1 N iodine solution (12.69 g iodine and 25 g KI made up to 1 L)
- 5% aqueous KOH (5 g KOH dissolved in H_2O and made up to 100 mL)

Experimental

- Place a drop of potato starch suspension on a microscope slide, cover with a coverslip, and examine under a microscope (100–600×) with

Food Carbohydrate Chemistry, First Edition. Ronald E. Wrolstad.
© 2012 John Wiley & Sons, Inc. Published 2012 by John Wiley & Sons, Inc.

ordinary light. Then examine under polarized light, rotating one of the filters so that crystallinity is observed by the presence of cross-striations. (Examination of potato starch is recommended because of its large granular size and distinctive shape.)

- Place a drop of I-KI solution at the edge of the coverslip and draw it into the starch sample with the aid of a piece of filter paper. Starch granules are blue or blue–black. (Cellulose appears yellow, and proteins brown or yellow.)
- Prepare another potato starch slide. Place a drop of KOH on the slide and draw it into the sample. KOH will disrupt hydrogen bonds and effectively gelatinize the starch. An increase in size will be observed. The swollen starch granules will no longer exhibit cross-striations under polarized light. This is an indication of loss of crystallinity.
- Examine other starch suspensions under the microscope, comparing their relative size, shape, appearance under polarized light, and color change with iodine.
- Examine a sample of potato starch that has been held at 60–65°C for a minimum of 1 hour, noting the color and appearance after treatment with iodine.

Note: Amylose will leach out of the granule when a starch suspension is held just below the gelatinization temperature (64°C for potato starch). Amylose will stain blue, and amylopectin red. If the demonstration is successful, the observed granule will be stained red, and the background field will be blue because of the amylose solution.

Anecdote

One evening many years ago, my daughter Greta was doing a middle-school science project at our dining room table. Her mother was taking an evening class, and I was doing the dishes. I had supplied her with a dropper bottle of iodine solution and numerous household items to test for presence of starch, e.g., corn starch, sugar, salt, tapioca, powdered sugar (contains 3% starch), Kleenex^{TM} tissue, notebook paper (coated with starch as a sizing agent), bread, slices of apple, banana, squash, etc. Midway through the project she announced, "Dad, I just spilled the bottle of iodine on the table, and I know one thing— our table isn't made of starch." I quickly dumped substantial quantities of cornstarch onto the iodine puddle. With successive applications, the cornstarch effectively removed the iodine stain from the antique oak table. There was no stain remaining by the time my wife Kathy returned.

Anecdote

In 1981, the Port of Morrow owned an industrial park in Boardman, OR, which included a potato processing plant. The Port (which leased the plant site to the potato processing plant) had the challenge of disposing of the large quantity of wastewater, which resulted from the potato-washing operation at the processing plant. The Port initially solved the disposal problem by converting the wastewater (with its nutrient enhancement of potato starch) into an asset by using it to irrigate some adjacent agricultural land. However, the problem that the Port encountered was that the impellers of the irrigation pumps were rapidly eroded by the high dirt content of the wastewater.

In order to reduce the dirt content in the wastewater, the Port contracted with Jensen Engineering to design a settlement pond. The idea was that by channeling the outflow of the wastewater into a settlement pond, the velocity of the outflow could be reduced to the point where the majority of the dirt would settle out.

However, the settlement pond did not produce the desired result. Instead, the quantity of dirt remained high and the damage to the pumps continued. The Port of Morrow filed a lawsuit against Jensen Engineering claiming that the settlement pond was defective.

During the pretrial investigation, it was discovered that the employees of the potato processing plant described the settlement pond as looking like "it was full of tapioca pudding." The wastewater coming into the settlement pond was observed to simply create its own direct path across the top of the "tapioca pudding"–filled settlement pond, flowing as a stream from the inlet at one end of the pond to the exit at the other end of the pond. Because this caused very little reduction in the velocity of the flow of the wastewater, this resulted in very little settlement of the dirt.

The key to the defense of the case was to determine the cause of the creation of the tapioca pudding in the settlement pond. Anyone who cooks knows that starch-based puddings and sauces will thicken during cooking. The cause of this is starch gelatinization. Although heat is required to gelatinize starch in most food systems, starch will gelatinize at ambient temperature at alkaline pH. The processing plant used lye peeling of potatoes so that the pH of the wastewater at times could be quite elevated. Potato starch has a swelling power of greater than 1000.

The jury decided that the cause of the problem in the settlement pond was not because of a defective design by Jensen Engineering, but rather was caused by the gelatinization of the starch in the wastewater. That cause was beyond the control of the defendant (W.A. Jerry North, Attorney, Schwabe, Williamson and Wyatt, Portland, OR, personal communication, 5/7/2011).

Reference

MacMasters MM. 1964. Microscopy. In: Whistler RL, Smith RJ, BeMiller JN, editors. *Methods in carbohydrate chemistry, vol IV.* New York: Academic Press, pp. 233–244.

Unit 9 Names and Structures of Oligosaccharides[a]

Appendix Table 9.1

Name[b]	Structure
Homogenous Reducing Oligosaccharides	
3-O-β-L-ara*f*-L-ara	
3-O-β-L-ara*p*-L-ara	
4-O-β-L-ara*p*-L-ara	

Food Carbohydrate Chemistry, First Edition. Ronald E. Wrolstad.
© 2012 John Wiley & Sons, Inc. Published 2012 by John Wiley & Sons, Inc.

Appendix Table 9.1   (*Continued*)

5-O-α-L-ara*f*-L-ara

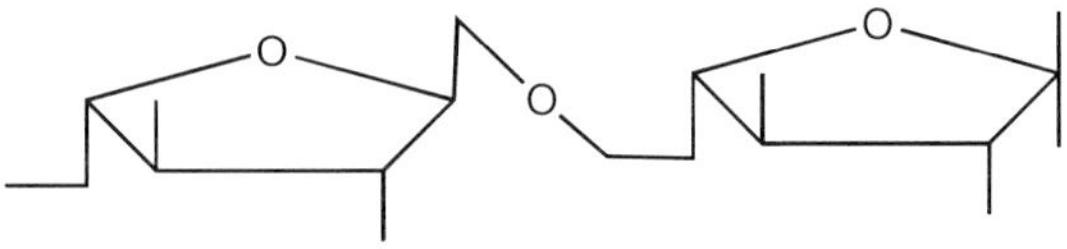

5-O-α-L-ara*p*-L-ara

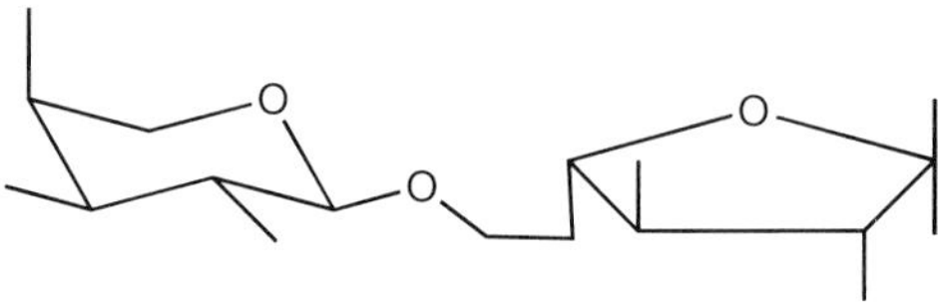

1-O-β-fru*f*-fru
(Inulobiose)

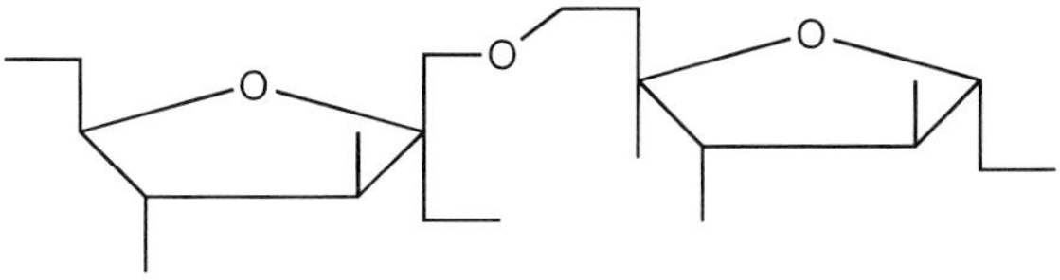

6-O-β-fru*f*-fru

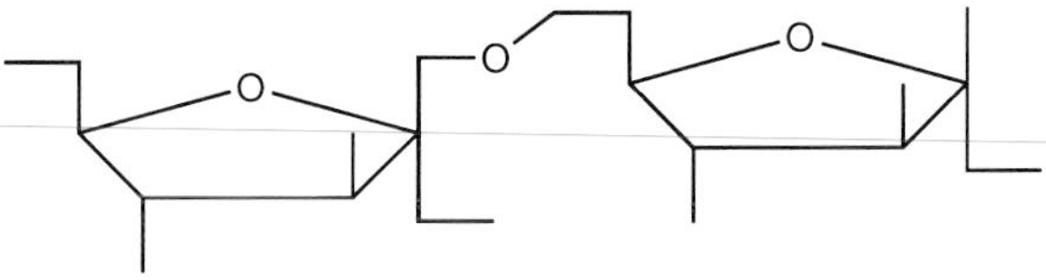

3-O-α-gal*p*-gal

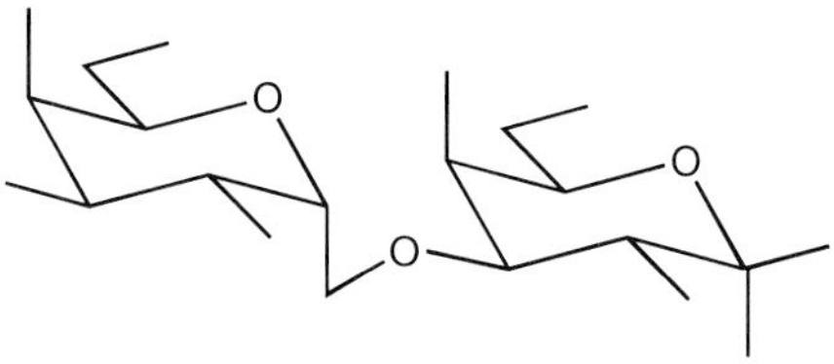

3-O-β-gal*p*-gal

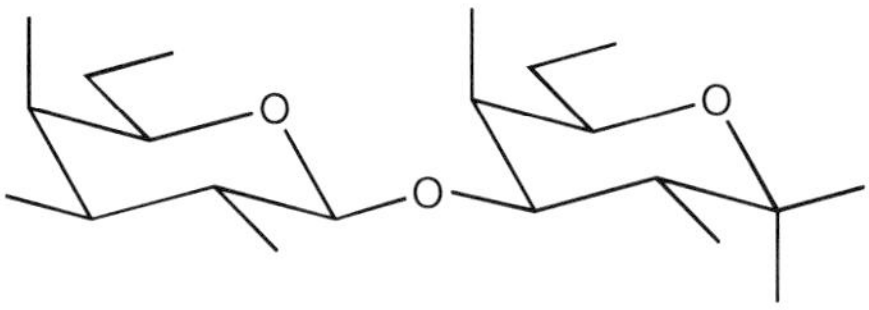

4-O-α-gal*p*-gal

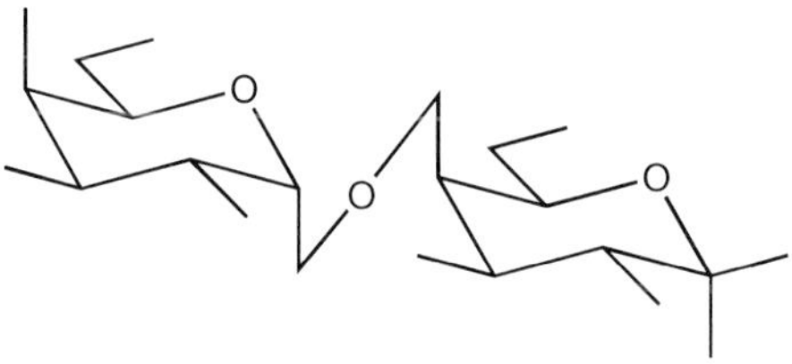

Appendix Table 9.1   (*Continued*)

4-O-β-gal*p*-gal (Galactobiose)	
5-O-α-gal*f*-gal*f*	
5-O-β-gal*f*-gal*f*	
6-O-α-gal*p*-gal	
6-O-β-gal*p*-gal	
2-O-α-glc*p*-glc (Kojibiose)	
2-O-β-glc*p*-glc (Sophorose)	

(*Continued*)

Appendix Table 9.1   (*Continued*)

3-O-α-glc*p*-glc (Nigerose)	
3-O-β-glc*p*-glc (Laminaribose)	
4-O-α-glc*p*-glc (Maltose)	
4-O-β-glc*p*-glc (Cellobiose)	
6-O-α-glc*p*-glc (Isomaltose)	
6-O-β-glc*p*-glc (Gentiobiose)	
2-O-α-man*p*-man	

Appendix Table 9.1   (*Continued*)

3-O-α-manp-man	
4-O-α-manp-man (Mannobiose)	
4-O-β-manp-man	
3-O-α-xylp-xyl	
3-O-β-xylp-xyl	
4-O-β-xylp-xyl (Xylobiose)	

Heterogeneous Reducing Oligosaccharides

2-O-β-fruf-glc	

(*Continued*)

Appendix Table 9.1   (*Continued*)

3-O-β-fru*f*-glc	
6-O-β-fru*f*-glc	
4-O-β-gal*p*-alt	
3-O-β-gal*p*-fru	
4-O-β-gal*p*-fru (Lactulose)	
6-O-β-gal*p*-fru	
2-O-α-gal*p*-glc	

Appendix Table 9.1   (*Continued*)

2-O-β-galp-glc	
3-O-α-galp-glc	
3-O-β-galp-glc	
4-O-β-galp-glc (Lactose)	
6-O-α-galp-glc (Melibiose)	
6-O-β-galp-glc (Allolactose)	
4-O-β-glap-man (Epilactose)	

(*Continued*)

Appendix Table 9.1   (*Continued*)

6-O-α-gal*p*-man	
4-O-β-glc*p*-alt (Celltrobiose)	
1-O-α-glc*p*-fru	
1-O-β-glc*p*-fru.2H$_2$O	
3-O-α-glc*p*-fru (Turanose)	
4-O-α-glc*p*-fru (Maltulose)	
4-O-β-glc*p*-fru (Cellobiulose)	
5-O-α-glc*p*-fru*p* (Leucrose)	

Appendix Table 9.1   (*Continued*)

6-O-α-glc*p*-fru (Isomaltulose)	
2-O-β-glc*p*-gal	
3-O-β-glc*p*-gal	
4-O-α-glc*p*-gal	
4-O-β-glc*p*-gal (Lycobiose)	
6-O-β-glc*p*-gal	
4-O-α-glc*p*-man (Epimaltose)	

Appendix Table 9.1 (*Continued*)

4-O-β-glc*p*-man (Epicellobiose)	
6-O-β-glc*p*-man	
4-O-β-gul*p*-glc	
4-O-β-man*p*-glc	
6-O-α-man*p*-glc	
6-O-β-man*p*-glc	
4-O-β-man*p*-L-gul	

Appendix Table 9.1 (*Continued*)

Homogeneous Nonreducing Disaccharides
α-ara*p* αara*p*
α-L-ara*p* α-L-ara*p*
β-L-ara*p* β-L-ara*p*
α-gal*p* α-gal*p*
α-gal*p* β-gal*p*
α-glc*p* α-glc*p* (α,α-Trehalose)
α-glc*p* β-glc*p* (Neotrehalose)

(*Continued*)

Appendix Table 9.1 (*Continued*)

β-glc*p* β-glc*p* (Isotrehalose)	
α-man*p* α-man*p*	
α-man*p* β-man*p*	
β-rib*f* β-rib*f*	
α-xyl*p* α-xyl*p*	

Heterogeneous Nonreducing Disaccharides

β-fru*f* α-gal*p* (Galsucrose)	
β-fru*f* α-glc*p* (Sucrose)	

Appendix Table 9.1   (*Continued*)

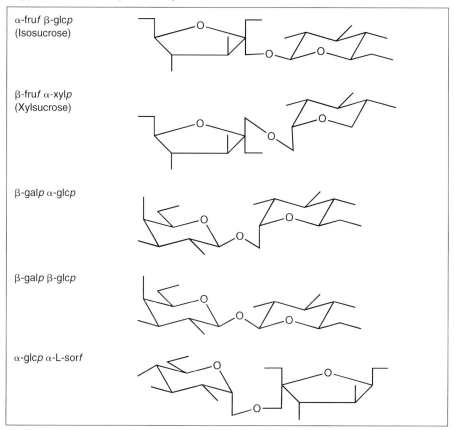

α-fru*f* β-glc*p*
(Isosucrose)

β-fru*f* α-xyl*p*
(Xylsucrose)

β-gal*p* α-glc*p*

β-gal*p* β-glc*p*

α-glc*p* α-L-sor*f*

[a]Modified from Shallenberger RS. 1982. *Advanced sugar chemistry*. Westport, CT: AVI Publishing, pp. 224–230.
[b]Abbreviations: Arabinose = ara; Altrose = alt; Fructose = fru; Galactose = gal; Glucose = glc; Gulose = gul; Mannose = man; Ribose = rib; Sorbose = sor; Xylose = xyl; Furanose = *f*; Pyranose = *p*.

Index

Food Carbohydrate Chemistry, First Edition. Ronald E. Wrolstad.
© 2012 John Wiley & Sons, Inc. Published 2012 by John Wiley & Sons, Inc.

214 Index

Food Science and Technology

WILEY-BLACKWELL

For further details and ordering information, please visit www.wiley.com/go/food

Food Science and Technology from Wiley-Blackwell

SENSORY SCIENCE, CONSUMER RESEARCH & NEW PRODUCT DEVELOPMENT

Sensory Evaluation: A Practical Handbook	Kemp	9781405162104
Statistical Methods for Food Science	Bower	9781405167642
Concept Research in Food Product Design and Development	Moskowitz	9780813824246
Sensory and Consumer Research in Food Product Design and Development	Moskowitz	9780813816326
Sensory Discrimination Tests and Measurements	Bi	9780813811116
Accelerating New Food Product Design and Development	Beckley	9780813808093
Handbook of Organic and Fair Trade Food Marketing	Wright	9781405150583
Multivariate and Probabilistic Analyses of Sensory Science Problems	Meullenet	9780813801780

FOOD LAWS & REGULATIONS

The BRC Global Standard for Food Safety: A Guide to a Successful Audit	Kill	9781405157964
Food Labeling Compliance Review, 4th edition	Summers	9780813821818
Guide to Food Laws and Regulations	Curtis	9780813819464
Regulation of Functional Foods and Nutraceuticals	Hasler	9780813811772

DAIRY FOODS

Dairy Ingredients for Food Processing	Chandan	9780813817460
Processed Cheeses and Analogues	Tamime	9781405186421
Technology of Cheesemaking, 2nd edition	Law	9781405182980
Dairy Fats and Related Products	Tamime	9781405150903
Bioactive Components in Milk and Dairy Products	Park	9780813819822
Milk Processing and Quality Management	Tamime	9781405145305
Dairy Powders and Concentrated Products	Tamime	9781405157643
Cleaning-in-Place: Dairy, Food and Beverage Operations	Tamime	9781405155038
Advanced Dairy Science and Technology	Britz	9781405136181
Dairy Processing and Quality Assurance	Chandan	9780813827568
Structure of Dairy Products	Tamime	9781405129756
Brined Cheeses	Tamime	9781405124607
Fermented Milks	Tamime	9780632064588
Manufacturing Yogurt and Fermented Milks	Chandan	9780813823041
Handbook of Milk of Non-Bovine Mammals	Park	9780813820514
Probiotic Dairy Products	Tamime	9781405121248

SEAFOOD, MEAT AND POULTRY

Handbook of Seafood Quality, Safety and Health Applications	Alasalvar	9781405180702
Fish Canning Handbook	Bratt	9781405180993
Fish Processing – Sustainability and New Opportunities	Hall	9781405190473
Fishery Products: Quality, safety and authenticity	Rehbein	9781405141628
Thermal Processing for Ready-to-Eat Meat Products	Knipe	9780813801483
Handbook of Meat Processing	Toldra	9780813821825
Handbook of Meat, Poultry and Seafood Quality	Nollet	9780813824468

BAKERY & CEREALS

Whole Grains and Health	Marquart	9780813807775
Gluten-Free Food Science and Technology	Gallagher	9781405159159
Baked Products – Science, Technology and Practice	Cauvain	9781405127028
Bakery Products: Science and Technology	Hui	9780813801872
Bakery Food Manufacture and Quality, 2nd edition	Cauvain	9781405176132

BEVERAGES & FERMENTED FOODS/BEVERAGES

Technology of Bottled Water, 3rd edition	Dege	9781405199322
Wine Flavour Chemistry, 2nd edition	Bakker	9781444330427
Wine Quality: Tasting and Selection	Grainger	9781405113663
Beverage Industry Microfiltration	Starbard	9780813812717
Handbook of Fermented Meat and Poultry	Toldra	9780813814773
Microbiology and Technology of Fermented Foods	Hutkins	9780813800189
Carbonated Soft Drinks	Steen	9781405134354
Brewing Yeast and Fermentation	Boulton	9781405152686
Food, Fermentation and Micro-organisms	Bamforth	9780632059874
Wine Production	Grainger	9781405113656
Chemistry and Technology of Soft Drinks and Fruit Juices, 2nd edition	Ashurst	9781405122863

PACKAGING

Food and Beverage Packaging Technology, 2nd edition	Coles	9781405189101
Food Packaging Engineering	Morris	9780813814797
Modified Atmosphere Packaging for Fresh-Cut Fruits and Vegetables	Brody	9780813812748
Packaging Research in Food Product Design and Development	Moskowitz	9780813812229
Packaging for Nonthermal Processing of Food	Han	9780813819440
Packaging Closures and Sealing Systems	Theobald	9781841273372
Modified Atmospheric Processing and Packaging of Fish	Otwell	9780813807683
Paper and Paperboard Packaging Technology	Kirwan	9781405125031

For further details and ordering information, please visit www.wiley.com/go/food